LECTURE NOTES
IN
NUMERICAL METHODS
OF
DIFFERENTIAL EQUATIONS

Professor Tadeusz STYŠ
Ph.D. in Mathematics
University of Warsaw

Warsaw, April 2009

CONTENTS

PREFACE

This text is intended for science and engineering students taking a course in numerical methods of differential equations. In addition to basic knowledge of mathematical analysis and differential equations, it requires some background in numerical analysis and computing. The book covers the following methods:

- Multistep linear difference methods

- Runge-Kutta methods

- Finite Difference Methods

Most of the material of this book has its origin based on lecture courses given to advanced undergraduate and early postgraduate students.
The `Mathematica` (cf. [33]) modulae are included to each chapter to illustrate the methods by graphical and numerical solutions and to use them in solving of application problems.

There is extensive literature published on numerical methods of differential equations including books which cover undergraduate and postgraduate programmes (cf. [[1],[2],[3],[6],[8],[9],[10],[12],[13],[14],[19],[20],[22],[24],[25],[27],[28]]. As the book for a course in numerical methods, this text surves the methods with use of `Mathematica` modulae. There is also desire that the reader will find verity of finite finite difference schemes and examples which help reading the text.

Two techniques of error analysis have been developed and used to prove convergence of the considered methods. The first technique is based on the discrete maximum principle and is applied to prove uniform convergence of the methods for linear and non-linear differential equations. The second technique draws on spectral analysis and deals with average convergence of the methods in Hilbert space H.

Chapter 1 concerns linear difference equations and it constitutes an introduction to linear multistep methods and finite difference methods.

Linear multistep methods and explicit Runge-Kutta methods are presented with solution of practical examples in chapter 2. Also, in chapter 2, the boundary value problem for he secand order of a system of ordinary differential equations is solved by an optimal algorithm with use of the included Mathematica module *solveBVP*.
Throughout the chapters 3, 4 and 5, the discrete maximum principle is given and applied to prove uniform convergence of finite difference schemes.

In chapter 3, considerable attention is paid to the construction of finite difference schemes and their global and local truncation errors.

Chapter 4 is designed for linear elliptic equations. Different variants of finite difference schemes are considered. The most interesting of these are finite difference schemes in canonical form. For such schemes the discrete maximum principle has been stated and used to estimate the global error of the method. Helmholtz equation is solved by `Mathematica` module `solveHelmholtz`.

Chapter 5 contains analysis of convergence of the finite difference scheme with weight for linear parabolic equations.

In chapter 6, the finite difference scheme with weight is applied to wave equation. These scheme has been solved by the method of separation of variables and by the module `waveEqn`.

Students are encourage to run `Mathematica` modulae, which are included in the tekst, to learn numerical methods by running the examples and solving tutorial questions with `Mathematica` as a system for Mathematics.

Tadeusz Styš   April, 2009

List of Mathematica Modulae

Chapter 1

Linear Difference Equations

Tadeusz Styš
University of Warsaw

Abstract: This chapter constitutes an introduction to linear multi step methods, Runge Kutta methods, and finite difference methods which are presented in the chapters that follow. The homogeneous and non-homogeneous linear difference equations are solved by the following *Mathematica* modulae: The module *differenceEqn* finds the general solution of a linear difference equation, the module *baseSolution* finds the fundamental set of solutions of a homogeneous linear difference equation, the module *particularSolution* finds a particular solution of a non-homogeneous linear difference equation by the method of variation of coefficients. The chapter ends with a set of questions.

Chapter 1

Linear Difference Equations

1.1   Homogeneous Linear Difference Equations

This chapter is an introduction to the finite difference methods for ordinary and partial differential equations. In fact, linear multi step methods, Runge-Kutta methods and finite difference methods lead to a special type of finite difference equations. Each of these equations will be separately considered in the following chapters.

We shall now turn to a general solution of homogeneous linear difference equation of order k

$$y_{n+k} + a_{k-1}y_{n+k-1} + \cdots + a_1 y_{n-1} + a_0 y_n = 0, \quad n = 0, 1, \ldots; \qquad (1.1)$$

where $a_0, a_1, \ldots, a_{k-1}$ are given real coefficients and $a_0 \neq 0$. Here, y_n, ($n = 0, 1, \ldots;$) is an unknown sequence.

Let us consider the *characteristic polynomial* of difference equation (1.1)

$$\rho(\xi) = \xi^k + a_{k-1}\xi^{k-1} + \ldots + a_1\xi + a_0. \qquad (1.2)$$

Substituting into (1.1)

$$y_{n+r} = \binom{r}{0}\Delta^0 y_n + \binom{r}{1}\Delta y_n + \binom{r}{2}\Delta^2 y_n + \cdots + \binom{r}{r}\Delta^r y_n, \quad r = 0, 1, \ldots, k,$$

we can rewrite equation (1.1) in terms of its characteristic polynomial $\rho(\xi)$ as follows:

$$\rho(1)y_n + \frac{\rho'(1)}{1!}\Delta y_n + \frac{\rho''(1)}{2!}\Delta^2 y_n + \cdots + \frac{\rho^{(k)}(1)}{k!}\Delta^k y_n = 0, \quad n = 0, 1, \ldots; \quad (1.3)$$

where the finite differences:

$$\Delta^0 y_n = y_n$$

$$\Delta y_n = y_{n+1} - y_n$$

$$\Delta^2 y_n = \Delta\Delta y_n = y_{n+2} - 2y_{n+1} + y_n$$

$$\cdots\cdots \quad \cdots\cdots\cdots\cdots\cdots\cdots\cdots\cdots\cdots\cdots\cdots\cdots\cdots$$

$$\Delta^k y_n = \Delta\Delta^{k-1} y_n = \binom{k}{0} y_{n+k} - \binom{k}{1} y_{n+k-1} + \cdots + (-1)^k \binom{k}{k} y_n,$$

and $\binom{k}{i} = \dfrac{k!}{i!\,(k-i)!}$, $i = 0, 1 \ldots, k$; are Newton's coefficients.

Each solution of difference equation (1.1) or (1.3) is determined by the roots of the characteristic equation

$$\rho(\xi) = \xi^k + a_{k-1}\xi^{k-1} + \ldots + a_1\xi + a_0 = 0.$$

In order to find all solutions of equation (1.1), suppose that

$$y_n = \lambda^n, \quad n = 0, 1, \ldots;$$

satisfies (1.1) for certain value of $\lambda \neq 0$, so that

$$\lambda^{n+k} + a_{k-1}\lambda^{n+k-1} + \cdots + a_1\lambda^{n+1} + a_0\lambda^n =$$

$$\lambda^n(\lambda^k + a_{k-1}\lambda^{k-1} + \cdots + a_1\lambda + a_0) =$$

$$\lambda^n \rho(\lambda) = 0.$$

Hence, we get the solution

$$y_n = \lambda_1^n, \quad n = 0, 1, \ldots;$$

of the homogeneous equation (1.1), provided that λ_1 is a root of the characteristic polynomial $\rho(\xi)$.

We shall be dealing with the following three cases:

1. All the roots of the characteristic polynomial $\rho(\xi)$ are real and distinct,

2. There are complex roots of the polynomial $\rho(\xi)$,

3. There are repeating roots of the polynomial $\rho(\xi)$.

Let us note that in case 1, where all the roots $\lambda_1, \lambda_2, \ldots, \lambda_k$ of the characteristic polynomial are real and distinct, we have the following k linearly independent real solutions of (1.1)

$$y_n^{(1)} = \lambda_1^n, \; y_n^{(2)} = \lambda_2^n, \; \ldots, \; y_n^{(k)} = \lambda_k^n, \quad n = 0, 1, \ldots; \tag{1.4}$$

Then, each solution of the homogeneous difference equation (1.1) can be represented as the linear combination of solutions (1.4), i.e.,

$$y_n = C_1 \lambda_1^n + C_2 \lambda_2^n + \cdots + C_k \lambda_k^n, \tag{1.5}$$

where $C_1, C_2, \ldots, C_k$ are constants uniquely determined by the initial values $y_0, y_1, \ldots, y_{k-1}$.

The solution y_n given by (1.5) is called the *general solution* of the homogeneous equation (1.1).

Example 1.1 *Find the general solution of the equation*

$$y_{n+2} - 5y_{n+1} + 6y_n = 0, \quad n = 0, 1, 2, \ldots; \tag{1.6}$$

and verify that this solution satisfies equation (1.6).

Solution. Clearly, the characteristic polynomial

$$\rho(\xi) = \xi^2 - 5\xi + 6 = 0$$

has two real and distinct roots $\lambda_1 = 2$ and $\lambda_2 = 3$. Therefore, the general solution of difference equation (1.6) is:

$$y_n = C_1 2^n + C_2 3^n, \quad n = 0, 1, \ldots;$$

where C_1 and C_2 are arbitrary constants.

Indeed, substituting this expression for y_n into equation (1.6) gives

$$y_{n+2} - 5y_{n+1} + y_n = C_1(2^{n+2} - 5 \; 2^{n+1} + 6 \; 2^n) + C_2(3^{n+2} - 5 \; 3^{n+1} + 6 \; 3^n) = 0,$$

for $n = 0, 1, \ldots$.

Example 1.2 *Find the general term of the Fibonacci sequence determined by the recursive formula*

$$y_{n+2} = y_{n+1} + y_n, \quad n = 0, 1, \ldots;$$

with the initial values

$$y_0 = 0 \quad \text{and} \quad y_1 = 1.$$

Solution. We shall find the general term of Fibonacci's sequence by solving the difference equation

$$y_{n+2} - y_{n+1} - y_n = 0, \quad n = 0, 1, \ldots;$$

under the initial conditions $y_0 = 0$ and $y_1 = 1$. The characteristic polynomial of this difference equation is

$$\rho(\xi) = \xi^2 - \xi - 1 = 0$$

and has two roots

$$\lambda_1 = \frac{1}{2}(1 - \sqrt{5}) \quad \text{and} \quad \lambda_2 = \frac{1}{2}(1 + \sqrt{5}).$$

Therefore, the general solution is

$$y_n = C_1[\frac{1 - \sqrt{5}}{2}]^n + C_2[\frac{1 + \sqrt{5}}{2}]^n, \quad n = 0, 1, \ldots;$$

where the constants C_1 and C_2 are determined by the initial conditions

$$y_0 = C_1 + C_2 = 0 \quad \text{and} \quad y_1 = C_1\frac{1 - \sqrt{5}}{2} + C_2\frac{1 + \sqrt{5}}{2} = 1.$$

Hence

$$C_1 = -\frac{1}{\sqrt{5}} \quad \text{and} \quad C_2 = \frac{1}{\sqrt{5}}.$$

Thus, the general term of Fibonacci's sequence is:

$$y_n = \frac{1}{\sqrt{5}}[(\frac{1 + \sqrt{5}}{2})^n - (\frac{1 - \sqrt{5}}{2})^n], \quad n = 0, 1, \ldots;$$

One can generate n terms of Fibonacci's sequence by the following `Mathematica` instructions with $n = 10$:

```
fibonaci[0]=0; fibonaci[1]=1;
fibonaci[n_]:=fibonaci[n]=fibonaci[n-1]+fibonaci[n-2];
Table[fibonaci[i],{i,0,10}]
```

output: {0, 1, 1, 2, 3, 5, 8, 13, 21, 34, 55}.

Now, let us consider the second case when the characteristic polynomial has a complex root. Let

$$\lambda = \mid \lambda \mid (\cos\psi + i\sin\psi)$$

be a complex root of the equation

$$\rho(\xi) = \xi^k + a_{k-1}\xi^{k-1} + \cdots + a_1\xi + a_0 = 0.$$

Then

$$y_n = \mid \lambda \mid^n (\cos n\psi + i\sin n\psi), \quad n = 0, 1, \ldots;$$

is a complex solution of the difference equation (1.1). Since the difference equation (1.1) is linear one, therefore both, the real part

$$Re\ y_n = \mid \lambda \mid^n \cos n\psi$$

and the imaginary part

$$Im\ y_n = \mid \lambda \mid^n \sin n\psi$$

are solutions of (1.1). Thus, each complex root $\lambda = \lambda_1$ produces two real and linearly independent solutions

$$y_n^{(1)} = \mid \lambda_1 \mid^n \cos n\psi \quad \text{and} \quad y_n^{(2)} = \mid \lambda_1 \mid^n \sin n\psi, \quad n = 0, 1, \ldots.$$

Example 1.3 *Find all the solutions of the following difference equation:*

$$y_{n+2} - y_{n+1} + y_n = 0, \quad n = 0, 1, \ldots;$$

Solution. Note that the characteristic polynomial

$$\rho(\xi) = \xi^2 - \xi + 1 = 0$$

has complex roots

$$\lambda_1 = \frac{1 + i\sqrt{3}}{2} \quad \text{and} \quad \lambda_2 = \frac{1 - i\sqrt{3}}{2}.$$

Since $\mid \lambda_1 \mid = \mid \lambda_2 \mid = 1$ and $Arg\lambda_1 = \psi = \frac{\pi}{3}$, there are two linearly independent solutions

$$y_n^{(1)} = \cos\frac{n\pi}{3} \quad \text{and} \quad y_n^{(2)} = \sin\frac{n\pi}{3}, \quad n = 0, 1, \ldots.$$

Therefore, the general solution is given by

$$y_n = C_1 \cos\frac{n\pi}{3} + C_2 \sin\frac{n\pi}{3}, \quad n = 0, 1, \ldots;$$

where C_1 and C_2 are arbitrary constants.

Finally, let us consider the third case, when the characteristic polynomial has a root of multiplicity $s \geq 2$. Below, we shall show that such root λ_1 generates the following s linearly independent solutions:

$$y_n^{(1)} = \lambda_1^n, \quad y_n^{(2)} = n\lambda_1^n, \quad \ldots \quad , y_n^{(s-1)} = n^{s-1}\lambda_1^n, \quad n = 0, 1, \ldots.$$

Let us find all the solutions generated by the root λ_1 in the following form:

$$y_n = \phi(n)\lambda_1^n, \quad n = 0, 1, \ldots;$$

where the sequence $\phi(n), \quad n = 0, 1, \ldots.$ is determined by the equation

$$\phi(n+k)\lambda_1^{n+k} + a_{k-1}\phi(n+k-1)\lambda_1^{n+k-1} + \cdots$$
$$+ a_1\phi(n+1)\lambda_1^{n+1} + a_0\phi(n)\lambda_1^n =$$
$$= \lambda_1^n[\phi(n+k)\lambda_1^k + a_{k-1}\phi(n+k-1)\lambda_1^{k-1} + \cdots$$
$$+ a_1\phi(n+1)\lambda_1 + a_0\phi(n)] = 0. \tag{1.7}$$

In order to find $\phi(n)$, $n = 0, 1, \ldots$, we can use the following identities:

$$\phi(n) \quad = \quad \phi(n)$$

$$\phi(n+1) \quad = \quad \phi(n) + \frac{\Delta\phi(n)}{1!}$$

$$\phi(n+2) \quad = \quad \phi(n) + \frac{\Delta\phi(n)}{1!} + \frac{\Delta^2\phi(n)}{2!}$$

$$\cdots\cdots\cdots \qquad \cdots\cdots\cdots\cdots\cdots\cdots\cdots\cdots\cdots\cdots$$

$$\phi(n+k-1) \quad = \quad \phi(n) + \frac{\Delta\phi(n)}{1!} + \frac{\Delta^2\phi(n)}{2!} + \cdots + \frac{\Delta^{k-1}\phi(n)}{(k-1)!}$$

$$\phi(n+k) \quad = \quad \phi(n) + \frac{\Delta\phi(n)}{1!} + \frac{\Delta^2\phi(n)}{2!} + \cdots + \frac{\Delta^{k-1}\phi(n)}{(k-1)!} + \frac{\Delta^k\phi(n)}{k!}$$

$$(1.8)$$

By substituting the identities (1.8) into equation (1.7) and collecting the coefficients of $\phi(n), \Delta\phi(n), \ldots, \Delta^k\phi(n)$, we obtain the following equation:

$$\phi(n+k)\lambda_1^k + a_{k-1}\phi(n+k-1)\lambda_1^{k-1} + \cdots$$

$$+a_1\phi(n+1)\lambda_1 + a_0\phi(n) \quad =$$

$$= \rho(\lambda_1)\phi(n) + \frac{\rho'(\lambda_1)}{1!}\lambda_1\Delta\phi(n) + \frac{\rho''(\lambda_1)}{2!}\lambda_1^2\Delta^2\phi(n) + \cdots$$

$$+\frac{\rho^{(k)}(\lambda_1)}{k!}\lambda_1^k\Delta^k\phi(n) \quad = \quad 0.$$

Since λ_1 is the root of the characteristic polynomial $\rho(\xi)$ of multiplicity s, we have

$$\rho(\lambda_1) = \rho'(\lambda_1) = \rho''(\lambda_1) = \cdots = \rho^{(s-1)}(\lambda_1) = 0, \quad \rho^{(s)}(\lambda_1) \neq 0,$$

and

$$\frac{\rho^{(s)}(\lambda_1)}{s!}\lambda_1^s\Delta^s\phi(n) + \frac{\rho^{(s+1)}(\lambda_1)}{(s+1)!}\lambda_1^{s+1}\Delta^{s+1}\phi(n) + \cdots + \frac{\rho^{(k)}(\lambda_1)}{k!}\lambda_1^k\Delta^k\phi(n) = 0.$$

The above equality holds if

$$\Delta^s\phi(n) = 0, \quad \Delta^{s+1}\phi(n) = 0, \quad \cdots \quad , \Delta^k\phi(n) = 0.$$

We can check that

$$\Delta^r\phi(n) = 0, \quad r = s, s+1, \ldots, k,$$

when

$$\phi(n) = n^q, \quad q = 0, 1, , \ldots, s-1.$$

Thus, the repeating root λ_1 yields s linearly independent solutions

$$y_n^{(q)} = n^q \lambda_1^n, \quad q = 0, 1, \ldots, s - 1.$$

Consequently, the general solution of (1.1) is:

$$y_n = \cdots + C_r \lambda_1^n + C_{r+1} n \lambda_1^n + C_{r+2} n^2 \lambda_1^n + \cdots + C_{r+s-1} n^{s-1} \lambda_1^n + \cdots \quad (1.9)$$

Example 1.4 *Find the general solution of the difference equation*

$$y_{n+3} - 6y_{n+2} + 12y_{n+1} - 8y_n = 0, \quad n = 0, 1, \ldots.$$

Solution. The characteristic polynomial

$$\rho(\xi) = \xi^3 - 6\xi^2 + 12\xi - 8 = 0$$

has a multiple root $\lambda_1 = 2$, with $s = 3$, which generates three linearly independent solutions

$$y_n^{(1)} = 2^n, \quad y_n^{(2)} = n2^n, \quad y_n^{(3)} = n^2 2^n, \quad n = 0, 1, \ldots.$$

Therefore, the general solution of this difference equation is:

$$y_n = C_1 \, 2^n + C_2 \, n \, 2^n + C_3 \, n^2 \, 2^n, \quad n = 0, 1. \ldots;$$

where C_1, C_2 and C_3 are arbitrary constants.
The general solution of a homogeneous linear difference equation can be obtain by the following `Mathematica` module:

Program 1.1 *Mathematica module for finding general solution of a homogeneous linear difference equations.*

```
differenceEqn[a_]:=Module[{b,l,l1,l2,m,i,p,r,u,
   s,co,generalsol},
   Clear[n];
   m=Length[a];
   p=Table[x^i,{i,0,m-1}];
   p=p.a;
   sol=Solve[p==0,x];
   sol=Flatten[sol];
   l=Table[sol[[i,2]],{i,1,m-1}];
   l1={l[[1]]};
   Do[If[!MemberQ[l1,l[[i]]],
   l1=Append[l1,l[[i]]]],{i,2,m-1}];
   l2={ };
```

```
Do[l2=Append[l2,{l1[[i]],
Length[Flatten[Position[l,l1[[i]],1]]]}],
{i,1,Length[l1]}];
r={ }; u={ };
Do[If[Im[l2[[i,1]]]===0,
r=Append[r,l2[[i]]],u=Append[u,l2[[i]]]],
{i,1,Length[l2]}];
b={ };
Do[b=Append[b,n^s*r[[i,1]]^n],
{i,1,Length[r]},{s,0,r[[i,2]]-1}];
Do[{b=Append[b,n^s Abs[u[[i,1]]]^n*
Cos[n Arg[u[[i,1]]]]],b=Append[b,
n^s Abs[u[[i,1]]]^n*Sin[n Arg[u[[i,1]]]]]},
{i,1,Length[u],2},{s,0,u[[i,2]]-1}];
co=Table[c[i],{i,1,Length[b]}];
generalsol=co.b
]
```

For example, to solve the linear difference equation

$$y_{n+4} - 3y_{n+3} + y_{n+2} + 4y_n = 0, \qquad n = 0, 1, ...,$$

we input data coefficients `a={4,0,1,-3,1}` and invoke the module `differenceEqn[a]`. Then, we obtain the following general solution:

$$y(n) = c_1 2^n + c_2 2^n \, n + c_3 \, Cos[\frac{2n\pi}{3}] + c_4 Sin[\frac{2n\pi}{3}]$$

for $n = 0, 1, ...$, where $c_1, c_2, c_3, c_4.$, are arbitrary coefficients.

1.2   Non-homogeneous Linear Difference Equations

The non-homogeneous linear difference equation

$$y_{n+k} + a_{k-1}y_{n+k-1} + \cdots + a_1 y_{n+1} + a_0 y_n = f_n, \qquad k = 0, 1, \ldots. \qquad (1.10)$$

can also be written in terms of its characteristic polynomial as follows:

$$\rho(1)y_n + \frac{\rho'(1)}{1!}\Delta y_n + \frac{\rho''(1)}{2!}\Delta^2 y_n + \cdots + \frac{\rho^{(k)}(1)}{k!}\Delta^k y_n = f_n, \qquad n = 0, 1, \ldots. \qquad (1.11)$$

This difference equation has the general solution

$$y_n = C_1 y_n^{(1)} + C_2 y_n^{(2)} + \cdots + C_k y_n^{(k)} + y_n^{(*)}, \qquad n = 0, 1, \ldots,$$

where $y_n^{(1)}, \; y_n^{(2)}, \; \ldots, y_n^{(k)}$ are the fundamental solutions of homogeneous equation (1.1) and $y_n^{(*)}$ is a particular solution of non-homogeneous equation (1.10).

Therefore, to solve a non-homogeneous difference equation, we find the general solution of the homogeneous difference equation, and then look for a particular solution of the non-homogeneous equation. Let us consider the simplest case where the right hand side $f_n = f$, $(n = 0, 1...)$ is a constant. By direct substitution, we can check that

$$
y_n^{(*)} = \begin{cases} \dfrac{f}{\rho(1)} & \text{if} \quad \rho(1) \neq 0, \\[3mm] n^s \dfrac{f}{\rho^{(s)}(1)} & \text{if} \quad \rho(1) = \rho'(1) = \cdots = \rho^{(s-1)}(1) = 0 \quad \text{and} \quad \rho^{(s)}(1) \neq 0. \end{cases}
$$

is a particular solution of equation (1.10).
In general, we can find a particular solution of equation (1.10) by the Lagrange's method of variation of parameters which is also successfully applicable to the difference linear equations with variable coefficients (cf. [16],[17]).

1.3   Method of variation of parameters

Let $y_n^{(1)}, y_n^{(2)}, \ldots, y_n^{(k)}$ be a fundamental set of solutions of homogeneous difference equation (1.1). We shall find a particular solution of non-homogeneous difference equation (1.10) in the following form:

$$
y_n^{(*)} = C_1(n)y_n^{(1)} + C_2(n)y_n^{(2)} + \cdots + C_k(n)y_n^{(k)}, \quad n = 0, 1, \ldots. \qquad (1.12)
$$

As there are k unknown coefficients $C_1(n), C_2(n), \ldots, C_k(n)$ to be determined, we can put, apart from equation (1.10), the additional $k - 1$ conditions

$$
\begin{aligned}
y_{n+1}^{(*)} &= C_1(n)y_{n+1}^{(1)} + C_2(n)y_{n+1}^{(2)} + \cdots + C_k(n)y_{n+1}^{(k)} \\[2mm]
y_{n+2}^{(*)} &= C_1(n)y_{n+2}^{(1)} + C_2(n)y_{n+2}^{(2)} + \cdots + C_k(n)y_{n+2}^{(k)} \\[2mm]
\ldots\ldots \quad \ldots \quad \ldots\ldots\ldots \quad \ldots \quad \ldots\ldots \quad \ldots\ldots\ldots \quad \ldots\ldots\ldots \\[2mm]
y_{n+k-1}^{(*)} &= C_1(n)y_{n+k-1}^{(1)} + C_2(n)y_{n+k-1}^{(2)} + \cdots + C_k(n)y_{n+k-1}^{(k)}
\end{aligned}
$$
$$(1.13)$$

Now, let us replace n by $n+1$ in (1.12) and subtract the first equation of (1.13) from (1.12). We then replace n by $n + 1$ in the first equation in (1.13), and subtract the second equation in (1.13). Replacing n by $n + 1$ in the second equation of (1.13) and subtract the third equation in (1.13). Proceeding in this way, we arrive at the following system of $k - 1$ equations:

$$\Delta C_1(n)y_{n+1}^{(1)} \quad + \quad \Delta C_2(n)y_{n+1}^{(2)} \quad + \cdots + \quad \Delta C_k(n)y_{n+1}^{(k)} \quad = 0$$

$$\Delta C_1(n)y_{n+2}^{(1)} \quad + \quad \Delta C_2(n)y_{n+2}^{(2)} \quad + \cdots + \quad \Delta C_k(n)y_{n+2}^{(k)} \quad = 0 \qquad (1.14)$$

$$\cdots\cdots\cdots \qquad \cdots\cdots\cdots \qquad \cdots \qquad \cdots\cdots\cdots \qquad \cdots$$

$$\Delta C_1(n)y_{n+k-1}^{(1)} \quad + \quad \Delta C_2(n)y_{n+k-1}^{(2)} \quad + \cdots + \quad \Delta C_k(n)y_{n+k-1}^{(k)} \quad = 0$$

We obtain one more equation substituting to (1.10) for $y_n^*, y_{n+1}^*, \dots, y_{n+k-1}^*$ right hand sides of (1.12), (1.13) and for

$$y_{n+k}^* = C_1(n+1)y_{n+k}^{(1)} + C_2(n+1)y_{n+k}^{(2)} + \cdots + C_k(n+1)y_{n+k}^{(k)}$$

Then, we get

$$\Delta C_1(n)y_{n+k}^{(1)} + \Delta C_2(n)y_{n+k}^{(2)} + \cdots + \Delta C_k(n)y_{n+k}^{(k)} = f_n, \quad n = 0, 1, \dots; \quad (1.15)$$

Hence, by the Cramer formulae, we obtain the following solution of the system of equations (1.14), (1.15):

$$\Delta C_s(n) = \frac{D_s(n)}{D(n)}, \quad s = 1, 2, \dots, k; \qquad (1.16)$$

where

$$D(n) = \begin{vmatrix} y_{n+1}^{(1)} & y_{n+1}^{(2)} & \cdots & y_{n+1}^{(k)} \\ y_{n+2}^{(1)} & y_{n+2}^{(2)} & \cdots & y_{n+2}^{(k)} \\ \cdots\cdots & \cdots\cdots & \cdots\cdots & \cdots\cdots \\ y_{n+k}^{(1)} & y_{n+k}^{(2)} & \cdots & y_{n+k}^{(k)} \end{vmatrix},$$

$$D_s(n) = \begin{vmatrix} y_{n+1}^{(1)} & y_{n+1}^{(2)} & \cdots & 0 & \cdots & y_{n+1}^{(k)} \\ y_{n+2}^{(1)} & y_{n+2}^{(2)} & \cdots & 0 & \cdots & y_{n+2}^{(k)} \\ \cdots\cdots & \cdots\cdots & \cdots & \cdots & \cdots & \cdots\cdots \\ y_{n+k}^{(1)} & y_{n+k}^{(2)} & \cdots & f_n & \cdots & y_{n+k}^{(k)} \end{vmatrix}, \quad s = 1, 2 \dots, k.$$

Let us write the first order difference equations (1.16) as follows:

$$C_s(1) - C_s(0) = \frac{D_s(0)}{D(0)}$$

$$C_s(2) - C_s(1) = \frac{D_s(1)}{D(1)}$$

$$C_s(3) - C_s(2) = \frac{D_s(2)}{D(2)}$$

$$\cdots\cdots\cdots\cdots\cdots\cdots$$

$$C_s(n) - C_s(n-1) = \frac{D_s(n-1)}{D(n-1)}$$

Adding both sides of the above equations, we get

$$C_s(n) - C_s(0) = \sum_{r=0}^{n-1} \frac{D_s(r)}{D(r)}, \quad s = 1, 2, \ldots, k.$$

Let $C_s(0) = 0$, $s = 1, 2, ..., k$, so that

$$C_s(n) = \sum_{r=0}^{n-1} \frac{D_s(r)}{D(r)}, \quad s = 1, 2, \ldots, k. \tag{1.17}$$

Hence, the particular solution of difference equation (1.10) is:

$$y_n^{(*)} = C_1(n)y_n^{(1)} + C_2(n)y_n^{(2)} + \cdots + C_k(n)y_n^{(k)}, \quad n = 0, 1, \ldots, \tag{1.18}$$

where the coefficients $C_1, C_2, \ldots, C_k$ are given by formulae (1.17).

Thus, to obtain a particular solution of a nonhomogeneous linear difference equation, first, we have to find a fundamental set of solutions of a homogeneous equation. Such a set of fundamental solution, we can determine using the following `Mathematica` module:

Program 1.2 *Mathematica module that finds the fundamental set of solutions of a homogeneous linear difference equation*

```
baseSolution[a_,n_]:=Module[{b,l,l1,l2,m,i,
        p,r,u,Des,co,sol},
        m=Length[a];
        p=Table[x^i,{i,0,m-1}];
        p=p.a;
        sol=Solve[p==0,x];
        sol=Flatten[sol];
        l=Table[sol[[i,2]],{i,1,m-1}];
        l1={l[[1]]};
        Do[If[!MemberQ[l1,l[[i]]],
        l1=Append[l1,l[[i]]]],{i,2,m-1}];
        l2={ };
        Do[l2=Append[l2,{l1[[i]],
        Length[Flatten[Position[l,l1[[i]],1]]]}],
        {i,1,Length[l1]}];
        r={ }; u={ };
        Do[If[Im[l2[[i,1]]]===0,
        r=Append[r,l2[[i]]],u=Append[u,l2[[i]]]],
        {i,1,Length[l2]}];
        b={ };
```

```
Do[b=Append[b,n^s*r[[i,1]]^n],
{i,1,Length[r]},{s,0,r[[i,2]]-1}];
Do[{b=Append[b,n^s Abs[u[[i,1]]]^n*
Cos[n Arg[u[[i,1]]]]],b=Append[b,
n^s Abs[u[[i,1]]]^n*Sin[n Arg[u[[i,1]]]]]},
{i,1,Length[u],2},{s,0,u[[i,2]]-1}];
b
]
```

For example, to find the fundamental set of solutions for the linear difference equation

$$y_{n+2} - 5y_{n+1} + 6y_n = 0, \quad n = 0, 1, ...,$$

we input data coefficients a={6,-5,1} and execute the command baseSolution[a,n]. Then, we obtain the following set of solutions:$\{2^n, 3^n\}$.

In order to find a particular solution of a nonhomogeneous linear difference equation, we can use the following module:

Program 1.3 *Mathematica module that finds a particular solution of a non-homogeneous linear difference equation when module* baseSolution *is active.*

```
particularSolution[a_,f_]:=Module[{d,dd,cs,i,m,s,c},
Clear[n,yn];
d[n_]:=Table[c[i]=baseSolution[a,n+i],
{i,1,Length[a]-1}];m=Length[d[1]];
flist=Table[0,{m-1}];
fl[n_]:=Append[flist,f[n]];
Do[dd[n_]:=Table[ReplacePart[Transpose[d[n]],fl[n],s],
{s,1,m}]];
cs[n_]:=Table[Sum[Det[dd[r][[s]]]/Det[d[r]],{r,0,n-1}],
{s,1,m}];base=baseSolution[a,n];
yn[n_]:=Sum[cs[n][[i]]*base[[i]],{i,1,m}];
yn[n]
]
```

For example, to find a particular solution of the equation

$$y_{n+2} - 5y_{n+1} + 6y_n = 6^n, \quad n = 0, 1, ...,$$

we input data coefficients a=6,-5,1 and define the function $f[n\_] := 6^n$ that appears in the right side. Then, invoking the module particularSolution[a], to obtain the particular solution

$$y(n) = \sum_{r=0}^{n-1} (3^{n-1}2^r - 2^{n-1}3^r).$$

Example 1.5 *Find the solution of the difference equation*

$$y_{n+2} - 5y_{n+1} + 6y_n = f_n, \quad n = 0, 1, \ldots; \tag{1.19}$$

which satisfies the initial conditions $y_0 = 0$ and $y_1 = 1$.

Solution. The characteristic equation

$$\rho(\xi) = \xi^2 - 5\xi + 6 = 0$$

has roots $\lambda_1 = 2$ and $\lambda_2 = 3$. Therefore, the fundamental set of solutions of the homogeneous equation is:

$$y_n^{(1)} = 2^n, \quad y_n^{(2)} = 3^n, \quad n = 0, 1, \ldots.$$

In order to find a particular solution of the non-homogeneous equation, we apply formula (1.18). Thus, we have

$$y_n^{(*)} = C_1(n)2^n + C_2(n)3^n, \quad n = 0, 1, \ldots,$$

where

$$C_1(n) = \sum_{r=0}^{n-1} \frac{D_1(r)}{D(r)}, \quad C_2(n) = \sum_{r=0}^{n-1} \frac{D_2(r)}{D(r)}, \quad n = 0, 1, \ldots.$$

Since

$$D(n) = \begin{vmatrix} 2^{n+1} & 3^{n+1} \\ 2^{n+2} & 3^{n+2} \end{vmatrix}, \quad D_1(n) = \begin{vmatrix} 0 & 3^{n+1} \\ f_n & 3^{n+2} \end{vmatrix}, \quad D_2(n) = \begin{vmatrix} 2^{n+1} & 0 \\ 2^{n+2} & f_n \end{vmatrix},$$

we obtain the coefficients

$$C_1(n) = -\sum_{r=0}^{n-1} \frac{f_r}{2^{r+1}}, \quad C_2(n) = \sum_{r=0}^{n-1} \frac{f_r}{3^{r+1}}, \quad n = 0, 1, \ldots;$$

and the particular solution

$$y_n^{(*)} = \sum_{r=0}^{n-1} f_r \left(\frac{1}{3^{r-n+1}} - \frac{1}{2^{r-n+1}} \right), \quad n = 0, 1, \ldots.$$

Hence, the general solution of the non-homogeneous (1.19) is:

$$y_n = C_1\, 2^n + C_2\, 3^n + \sum_{r=0}^{n-1} f_r \left(\frac{1}{3^{r-n+1}} - \frac{1}{2^{r-n+1}} \right), \quad n = 0, 1, \ldots,$$

where constants C_1 and C_2 are determined by the initial conditions

$$y_0 = C_1 + C_2 = 0, \quad y_1 = 2C_1 + 3C_2 = 1.$$

Thus, the solution of the initial problem (1.19) is:

$$y_n = 3^n - 2^n + \sum_{r=0}^{n-1} f_r \left(\frac{1}{3^{r-n+1}} - \frac{1}{2^{r-n+1}} \right), \quad n = 0, 1, \ldots;$$

1.4   Exercises

Question 1.1 *Find the general solution of the following equations:*

1. $y_{n+2} - 3y_{n+1} - 4y_n = 0, \quad n = 0, 1, \ldots;$

2. $y_{n+3} + 3y_{n+2} + 4y_n = 0; \quad n = 0, 1, \ldots;$

3. $y_{n+2} - 2y_{n+1} + 2y_n = 0, \quad n = 0, 1, \ldots;$

Question 1.2 *Find a particular solution of the equations*

1. $y_{n+2} - 5y_{n+1} + 4y_n = n, \quad n = 0, 1, \ldots;$

2. $y_{n+2} - 6y_{n+1} + 8y_n = 6, \quad n = 0, 1, \ldots;$

Question 1.3 *Solve the following equations with indicated initial value conditions:*

1. $3y_{n+2} - 5y_{n+1} + 2y_n = 0, \quad n = 0, 1, \ldots; \quad y_0 = 1, \ y_1 = 0,$

2. $y_{n+2} - 4y_{n+1} + 5y_n = 2, \quad n = 0, 1, \ldots; \quad y_0 = 0, \ y_1 = 2.$

Question 1.4 *Use the Lagrange's method of variation of parameters to find a solution of the equation*

$$y_{n+2} - 6y_{n+1} + 5y_n = \frac{1}{n+1}, \quad n = 0, 1, \ldots;$$

which satisfies the initial conditions $y_0 = 1, \quad y_1 = 0.$

Question 1.5 *Consider the following linear difference equation:*

$$y_{n+2} - \alpha \, y_{n+1} + y_n = 0, \quad n = 0, 1, \ldots; \qquad (1.20)$$

1. *Find the range of parameter α for which all solutions of equation (1.20) are bounded.*

2. *Give an estimate of the solution y_n, $(n = 0, 1, \ldots,)$ when $y_0 = 0$, and $y_1 = 1$.*

Question 1.6 *Use* Mathematica *module* differenceEqn *to find the general solution of the equations*

1. *(a)*
$$y_{n+4} - 3y_{n+3} + y_{n+2} + 4y_n = 0, \quad n = 0, 1, \ldots,$$

   (b)
$$y_{n+3} - 3y_{n+2} + 3y_{n+1} - y_n = 0, \quad n = 0, 1, \ldots$$

Question 1.7 *Find a set of fundamental solutions of the equation*

$$y_{n+7} - 5y_{n+6} + 12y_{n+5} - 18y_{n+4} + 18y_{n+3} - 12y_{n+2} + 5y_{n+1} - y_n = 0, \quad n = 0, 1, \ldots$$

using the module `baseSolution`

Question 1.8 *Find a particular solution of the equation*

$$y_{n+2} - 8y_{n+1} + 15y_n = 15^n, \quad n = 0, 1, \ldots$$

using the module `particularSolution`

Chapter 2

Solution of Ordinary Differential Equations

Tadeusz Styš
University of Warsaw

Abstract: In this chapter, the linear multi step methods and Runge Kutta methods are considered. For both classes of methods, efficient algorithms have been built and implemented in *Mathematica*. The designed *Mathematica* modulae are applied to selected initial value problems for ordinary differential equations and system of equations of the first order. Also, in chapter 2, an optimal algorithm has been built and implemented in the *Mathematica* module *solveBVP*. The module finds the numerical solution $u = (u_1, u_2, , up)$ of the system of equations

$$-\frac{d^2u}{dx^2} = f(x, u), \qquad a < x < b$$

with the boundary value conditions $u(a) = a_0$, $u(b) = b_0$. As exercises, the equations on rotation of a heavy string and on equilibrium of a rod are solved by the module. The chapter ends with a set of questions.

Chapter 2

Solution of Ordinary Differential Equations

2.1   Introduction

There is a great deal of literature concerning numerical methods for solving ordinary differential equations. This vast literature has been published in numerous articles and books, (cf. [8],[10],[13],[14]). We shall concentrate on linear multistep methods and explicit Runge-Kutta methods. These two well established methods are applicable to systems of ordinary differential equations. In order to simplify notation, we shall present these methods for solving the following equation:

$$\frac{dy(x)}{dx} = f(x, y(x)), \quad a \leq x \leq b, \tag{2.1}$$

with the initial value condition

$$y(a) = y_a. \tag{2.2}$$

Throughout the chapter, we shall assume that $f(x, y)$ is a continuous function which satisfies the Lipschitz condition with respect to the variable y, that is, there exists a constant L (Lipschitz constant) such that

$$|f(x, y) - f(x, z)| \leq L\, |y - z|$$

for all $x \in [a, b]$, and $y, z \in [c, d]$.
The following Cauchy-Picard theorem holds:

Theorem 2.1 *If $f(x, y)$ is a continuous function which satisfies the Lipschitz condition with respect to variable y then the initial value problem (2.1), (2.2) has a unique regular solution $y(x)$ in interval $[a, b]$.*

2.2   Linear Multistep Methods

All linear multistep methods can be written in the following general form:

$$\alpha_0 y_n + \alpha_1 y_{n+1} + \cdots + \alpha_{k-1} y_{k-1} + y_{n+k} = h[\beta_0 f_n + \beta_1 f_{n+1} + \cdots + \beta_k y_{n+k}], \quad (2.3)$$

or in sigma notation

$$\sum_{j=0}^{k} \alpha_j y_{n+j} = h \sum_{j=0}^{k} \beta_j f_{n+j}, \quad n = 0, 1, \ldots, N - k, \qquad (2.4)$$

where

$$x_i = a + ih, \quad i = 0, 1, \ldots, N; \quad h = \frac{b - a}{N},$$

$$y(x_n) \; - \; the \; value \; of \; the \; theoretical \; solution \; y(x) \; at \; x_n,$$

$$y_n \; - \; an \; approximate \; value \; of \; y(x_n),$$

$$f_n \; - \; the \; value \; of \; f(x, y) \; at \; the \; point \; (x_n, y_n).$$

We shall assume that $\alpha_k = 1$ and $\alpha_0^2 + \beta_0^2 > 0$. The assumption $\alpha_0^2 + \beta_0^2 > 0$ does not restrict the class of all linear multistep methods, since then the number of steps k can be reduced to $k - 1$, when $\alpha_0 = \beta_0 = 0$.
Each method of class (2.3) is called *an implicit method* if the coefficient $\beta_k \neq 0$, otherwise it is referred to as *an explicit method*.
In order to determine the $\alpha's$ and $\beta's$ coefficients, we introduce the following linear operator:

$$P[y](x) = \sum_{j=0}^{k} [\alpha_j y(x + jh) - h\beta_j y'(x + jh)] \qquad (2.5)$$

associated with the general form (2.4) of the linear multistep methods. Evidently, this operator is well defined on the set of all differentiable functions $y(x)$ in the interval $[a, b]$. Assuming that $y(x)$ is sufficiently smooth, i.e., $y(x)$ has as many derivatives as required, we can present operator P in the following form:

$$P[y](x) = C_0 y(x) + C_1 h y'(x) + C_2 h^2 y''(x) + \cdots + C_q h^q y^{(q)}(x) + \cdots \qquad (2.6)$$

where

$$C_0 = \alpha_0 + \alpha_1 + \cdots + \alpha_{k-1} + 1,$$

$$C_1 = (\alpha_1 + 2\alpha_2 + 3\alpha_3 + \cdots + (k-1)\alpha_{k-1} + k) - (\beta_0 + \beta_1 + \cdots + \beta_k),$$

$$C_2 = \frac{1}{2!}(\alpha_1 + 2^2\alpha_2 + 3^2\alpha_3 + \cdots + (k-1)^2\alpha_{k-1} + k^2)$$
$$- (\beta_1 + 2\beta_2 + 3\beta_3 + \cdots + k\beta_k),$$

$$\cdots \cdots \cdots \cdots \cdots \cdots \cdots \cdots \cdots \cdots \cdots \cdots \cdots \cdots \cdots \cdots \cdots$$

$$C_q = \frac{1}{q!}(\alpha_1 + 2^q\alpha_2 + 3^q\alpha_3 + \cdots + (k-1)^q\alpha_{k-1} + k^q)$$
$$- \frac{1}{(q-1)!}(\beta_1 + 2^{q-1}\beta_2 + 3^{q-1}\beta_3 + \cdots + k^{q-1}\beta_k),$$

for $q = 2, 3, \ldots$;

$$(2.7)$$

To show that formula(2.7) holds, let us substitute into (2.6)

$$y(x + jh) = y(x) + \frac{y'(x)}{1!}jh + \frac{y''(x)}{2!}(jh)^2 + \cdots + \frac{y^{(q)}(x)}{q!}(jh)^q + \cdots$$

$$y'(x + jh) = y'(x) + \frac{y''(x)}{1!}jh + \frac{y'''(x)}{2!}(jh)^2 + \cdots + \frac{y^{(q+1)}(x)}{q!}(jh)^q + \cdots$$

Then, we get

$$P[y](x) = \sum_{j=0}^{k}\{\alpha_j[y(x) + \frac{y'(x)}{1!}jh + \frac{y''(x)}{2!}(jh)^2 + \cdots + \frac{y^{(q)}(x)}{q!}(jh)^q + \cdots]$$

$$- \beta_j h[y'(x) + \frac{y''(x)}{1!}jh + \frac{y'''(x)}{2!}(jh)^2 + \cdots + \frac{y^{(q+1)}(x)}{q!}(jh)^q + \cdots]\}$$

$$= \sum_{j=0}^{k}\alpha_j y(x) + \sum_{j=0}^{k}j(\alpha_j - \beta_j)hy'(x) + \sum_{j=1}^{k}(\frac{j^2\alpha_j}{2!} - j\beta_j)h^2y''(x)$$

$$+ \sum_{j=1}^{k}(\frac{j^3\alpha_j}{3!} - \frac{j^2\beta_j}{2!})h^3y'''(x) + \cdots + \sum_{j=1}^{k}(\frac{j^q\alpha_j}{q!} - \frac{j^{q-1}\beta_j}{(q-1)!})h^qy^{(q)} + \cdots$$

$$= C_0y(x) + C_1hy'(x) + C_2h^2y''(x) + \cdots + C_qh^qy^{(q)}(x) + \cdots$$

We note that the coefficients at $h^sy^{(s)}$ for $s = 0, 1, ..., q$, in the above equality are given by (2.7).

2.2.1   Order of a method and local truncation error

There is a relationship between order of a linear multistep method and its local truncation error. We shall establish this relationship using formula (2.6) and the following definition:

Definition 2.1 *A linear multistep method (or a difference operator P) is said to be of the order p if and only if*

$$C_0 = 0, \ C_1 = 0, \ \ldots, C_p = 0 \quad and \quad C_{p+1} \neq 0, \tag{2.8}$$

where $C_0, \ C_1, \ldots, C_{p+1}$ are the coefficients of operator P given by formulas (2.7).

Clearly, a linear operator P of the order p takes the following forms:

$$P[y](x) = C_{p+1}h^{p+1}y^{(p+1)}(x) + \cdots \tag{2.9}$$

or

$$P[y](x) = C_{p+1}h^{p+1}y^{(p+1)}(\xi_x), \tag{2.10}$$

for a certain point $\xi_x \in (a, b)$.
The quantity

$$C_{p+1}h^{p+1}y^{(p+1)}(\xi_x)$$

is called *the local truncation error*, and C_{p+1} is called the *error constant*.
Below, we shall consider particular cases of the general form (2.3).

2.2.2   Linear one step methods

Setting $k = 1$ in the general form of linear multistep methods (2.1), we obtain the subclass

$$\alpha_0 y_n + y_{n+1} = h[\beta_0 f_n + \beta_1 f_{n+1}], \quad n = 0, 1, \ldots, N - 1; \tag{2.11}$$

of one-step methods, where $y_0 = y_a$ is a given starting value.
In (2.11), there are three free coefficients α_0, β_0 and β_1 to be determined. We shall find these coefficients assuming the order $p \geq 1$. By conditions (2.8), we have

$$C_0 = \alpha_0 + 1 = 0, \qquad \alpha_0 = -1,$$
$$C_1 = 1 - (\beta_0 + \beta_2) = 0, \quad \beta_0 + \beta_1 = 1. \tag{2.12}$$

This system of two equations has an infinite number of solutions. Let us choose some of them to obtain the known one step methods.
The Implicit Euler Rule. Let $\alpha_0 = -1$, $\beta_0 = 0$ and $\beta_1 = 1$. Then, we get the implicit Euler method

$$y_{n+1} = y_n + h \, f_{n+1}, \quad n = 0, 1, \ldots, N - 1, \tag{2.13}$$

with the initial condition $y_0 = y_a$.
The implicit Euler method has order $p = 1$ and the local truncation error

$$P[y](x) = C_2 h^2 y''(\xi_x), \quad \xi_x \in (a, b),$$

where the error constant $C_2 = -\frac{1}{2}$.

The Explicit Euler Rule. Let $\alpha_0 = -1$, $\beta_0 = 1$ and $\beta_1 = 0$. Then, we get the explicit Euler method

$$y_{n+1} = y_n + h\, f_n, \quad n = 0, 1, \ldots, N - 1, \tag{2.14}$$

with the initial condition $y_0 = y_a$. The explicit Euler method has also order $p = 1$ and its local truncation error is

$$P[y](x) = C_2 h^2 y''(\xi_x), \quad \xi_x \in (a, b),$$

where the error constant $C_2 = \frac{1}{2}$.

The Trapezian Rule. Let $\alpha_0 = -1$, $\beta_0 = \frac{1}{2}$ and $\beta_1 = \frac{1}{2}$. Then, we get the trapezian method

$$y_{n+1} = y_n + \frac{h}{2}[f_n + f_{n+1}], \quad n = 0, 1, \ldots, N - 1, \tag{2.15}$$

with the initial condition $y_0 = y_a$.

The trapezian method has order $p = 2$, since

$$C_0 = 0, \; C_1 = 0, \quad C_2 = 0, \quad C_3 = \frac{1}{3!} - \frac{1}{2!}\beta_1 = -\frac{1}{12}.$$

Therefore, its local truncation error is

$$P[y](x) = C_3 h^3 y''(\xi_x), \quad \xi_x \in (a, b),$$

where the error constant $C_3 = -\frac{1}{12}$.

Let us note that among all one step methods only trapezian method has order $p = 2$, i.e., there is no a method of order greater than 2 within the subclass (2.11).

2.2.3 Linear two step methods

Setting $k = 2$ in the general form of the linear multistep methods (2.1), we obtain the following class of linear two-step methods:

$$\alpha_0 y_n + \alpha_1 y_{n+1} + y_{n+2} = h[\beta_0 f_n + \beta_1 f_{n+1} + \beta_2 f_{n+2}], \quad n = 0, 1, \ldots, N-2. \tag{2.16}$$

In order to start with a two-step method, we need two starting values while one of them $y_0 = y_a$ is given in the initial condition, the another one y_1 should be evaluated by a suitable method.

The general form (2.16) of linear two-step methods contains five unknown parameters $\alpha_0, \alpha_1, \beta_0, \beta_1$ and β_2. To determine these parameters, we can assume the order of a method required. Thus, assuming that we are looking for a linear two-step method of order $p = 2$, we shall arrive to the following conditions:

$$C_0 = \alpha_0 + \alpha_1 + 1 = 0,$$

$$C_1 = (\alpha_1 + 2) - (\beta_0 + \beta_1 + \beta_2) = 0,$$

$$C_2 = \tfrac{1}{2}(\alpha_1 + 4) - (\beta_1 + 2\beta_2) = 0,$$

$$C_3 = \tfrac{1}{6}(\alpha_1 + 8) - \tfrac{1}{2}(\beta_1 + 4\beta_2) \neq 0.$$

$$(2.17)$$

2.2.4 The Mid-Point Rule

Setting in (2.17) $\alpha_0 = -1$, $\alpha_1 = 0$, $\beta_0 = 0$, $\beta_1 = 2$ and $\beta_2 = 0$, we get the mid-point method

$$y_{n+2} = y_n + 2hf_{n+1}, \quad n = 0, 1, \ldots, N - 2. \qquad (2.18)$$

To use this method, we need two starting values $y_0 = y_a$ and y_1. The value $y_0 = y_a$ is given in the initial condition, and y_1 is to be calculated by a suitable one-step method.

The mid-point method has the local truncation error

$$P[y](x) = C_3 h^3 y'''(\xi_x), \quad \xi_x \in (a, b),$$

where the error constant $C_3 = \dfrac{1}{3}$.

Now, let us choose a linear two-step method of the highest order. For the five parameter's $\alpha_0, \alpha_1, \beta_0, \beta_1$ and β_2, we can put the following five conditions:

$$C_0 = \alpha_0 + \alpha_1 + 1 = 0,$$

$$C_1 = (\alpha_1 + 1) - (\beta_0 + \beta_1 + \beta_2) = 0,$$

$$C_2 = \tfrac{1}{2}(\alpha_1 + 4) - (\beta_1 + 2\beta_2) = 0,$$

$$C_3 = \tfrac{1}{6}(\alpha_1 + 8) - \tfrac{1}{2}(\beta_1 + 4\beta_2) = 0,$$

$$C_4 = \tfrac{1}{24}(\alpha_1 + 16) - \tfrac{1}{6}(\beta_1 + 8\beta_2) = 0.$$

$$(2.19)$$

Simpson's Rule. Setting in (2.19) $\alpha_0 = -1$, $\alpha_1 = 0$, $\beta_0 = \dfrac{1}{3}$, $\beta_1 = \dfrac{4}{3}$ and $\beta_2 = \dfrac{1}{3}$, we obtain the Simpson's method

$$y_{n+2} = y_n + \frac{h}{3}(f_n + 4f_{n+1} + f_{n+2}), \quad n = 0, 1, \ldots, N - 2. \qquad (2.20)$$

To use the Simpson's method, we need two starting values y_0 and y_1. The value $y_0 = y_a$, and the value y_1 is to be determined by a suitable method. The Simpson's method has highest order $p = 4$ among all linear two step methods and its local truncation error is given by

$$P[y](x) = C_5 h^5 \, y^{(5)}(\xi_x), \quad \xi_x \in (a, b),$$

where the error constant $C_5 = -\dfrac{1}{90}$.

In order to find a method of possibly highest order within the class of k-step methods, we can use the following module:

Program 2.1 *Mathematica module that generates the method of the highest order within the class of k-step methods*

```
multistepMethod[k_]:=Module[{a, be, f, i, j, m, q, sol},
m=2 k+1;
a=Table[0,{i,1,m},{j,1,m}];
Do[If[i<=k,a[[1,i]]=1,0],{i,1,m}];
Do[If[i<=k,a[[2,i]]=i-1,a[[2,i]]=-1],{i,1,m}];
Do[If[i<=k,a[[q,i]]=(i-1)^(q-1)/(q-1)!,
a[[q,i]]=-(i-k-1)^(q-2)/(q-2)!],{i,2,m},{q,3,m}];
f=Table[-k^i/i!,{i,0,m-1}];
sol=LinearSolve[a,f];
sol=Insert[sol,1,k+1];
al=Take[sol,k+1]; be=Take[sol,-(k+1)];
sol={al,be}
]
```

For example, to obtain Simpson's method, we execute the following `Mathematica` command: `multistepMethod[2]`, to obtain the list of coefficients

$$\{\{-1,0,1\},\{1/3,4/3,1/3\}\},$$

where $(\alpha_0, \alpha_1, \alpha_2) = (-1, 0, 1)$ and $(\beta_0, \beta_1, \beta_2) = (\dfrac{1}{3}, \dfrac{4}{3}, \dfrac{1}{3})$.

Consistency. A linear multistep method of class (2.3) should be consistent with the differential equation (2.1). We can express the consistency of a method in terms of its order, or equivalently in terms of its characteristic polynomials. Namely, the following definition is used:

Definition 2.2 *(cf. [14], p.30) A linear multistep method is said to be consistent if it has order $p \geq 1$.*

All the one-step and two-step methods derived above have an order of at least $p = 1$, and therefore they are consistent with the differential equation (2.1). Let us note that a linear multistep method is consistent if the coefficients $C_0 = 0$ and $C_1 = 0$, since then $p \geq 1$. We can express these conditions in terms of the following characteristic polynomials:

$$\rho(\xi) = \xi^k + \alpha_{k-1}\xi^{k-1} + \cdots + \alpha_1\xi + \alpha_0,$$

$$\sigma(\xi) = \beta_k\xi^k + \beta_{k-1}\xi^{k-1} + \cdots + \beta_\xi + \beta_0.$$

Indeed, we have

$$C_0 = \alpha_0 + \alpha_1 + \cdots + \alpha_{k-1} + 1 = \rho(1)$$

and

$$C_1 = (\alpha_1 + 2\alpha_2 + \cdots + (k-1)\alpha_{k-1} + k\alpha_k) - (\beta_0 + \beta_1 + \cdots + \beta_k) = \rho'(1) - \sigma(1).$$

Thus, we have arrived at the following necessary and sufficient conditions for consistency of a linear multistep method:

$$\rho(1) = 0, \qquad \rho'(1) = \sigma(1). \tag{2.21}$$

Example 2.1 *The linear two-step method*

$$y_{n+2} + 4y_{n+1} - 5y_n = 2h(2f_{n+2} + f_n), \quad n = 0, 1, \ldots, N - 2,$$

is consistent, since it satisfies conditions (2.21).

Indeed, its first characteristic polynomial

$$\rho(\xi) = \xi^2 + 4\xi - 5$$

and its second characteristic polynomial

$$\sigma(\xi) = 4\xi^2 + 2.$$

Hence $\rho(1) = 0$ and $\rho'(1) = \sigma(1) = 6$.

2.2.5   Stability

We shall deal with stability in the sense of the following definition:

Definition 2.3 *A linear multistep method is said to be stable if the homogeneous difference equation*

$$y_{n+k} + \alpha_{k-1}y_{n+k-1} + \alpha_{k-2}y_{k-2} + \cdots + \alpha_1y_{n+1} + \alpha_0y_n = 0, \quad n = 0, 1, \ldots, N - k,$$

has all solutions bounded.

Let us note that all solutions of a homogeneous finite difference equation are bounded if and only if the characteristic polynomial $\rho(\xi)$ has all its roots within the unit circle and those with modulus one are simple.

Indeed, each solution of a homogeneous linear difference equation takes the following form (1.9):

$$y_n = C_1 y_n^{(1)} + C_2 y_n^{(2)} + \cdots + C_k y_n^{(k)}.$$

Therefore, the solution y_n, $n = 0, 1, ...$, is bounded if the fundamental solutions $y_n^{(r)} = n^{s-1} \lambda_r^n$, $r = 1, 2, ..., k$ are bounded, i.e. there exists a constant C such that

$$|y_n^{(r)}| = |n^{s-1} \lambda_r^n| \leq C$$

for any multiplicity s of the root λ_r and for all $r = 1, 2, ..., k$, $n = 0, 1, ...$. Hence

$$|n^{s-1} \lambda_r^n| \leq C, \quad n = 0, 1, ...$$

provided that either $|\lambda_r| < 1$ or $|\lambda_r| = 1$ and $s = 1$. Thus, definition (2.3) is equivalent to the following:

Definition 2.4 *A linear multistep method is said to be stable if no root of the first characteristic polynomial $\rho(\xi)$ has modulus greater than one, and every root with modulus one is simple.*

-

The implicit and explicit Euler methods and the trapezian method are stable since their first characteristic polynomial $\rho(\xi) = \xi - 1$ has only one root $\xi_1 = 1$ in the unit circle.

Within the class of two step methods, both the mid-point method and the Simpson's method are stable, since their first characteristic polynomial $\rho(\xi) = \xi^2 - 1$ has distinct roots $\xi_1 = 1$ and $\xi_2 = -1$ in the unit circle.

Example 2.2 *Consider the following linear two-step method:*

$$y_{n+2} + a y_{n+1} + b y_n = h(2 f_{n+1} + a f_n), \quad n = 0, 1, \ldots, N - 2.$$

Find the range of parameters a and b for which the method is consistent and stable. For which values of a and b does the method have order $p = 2$ and is consistent and stable ?

The characteristic polynomials

$$\rho(\xi) = \xi^2 + a\xi + b \quad \text{and} \quad \sigma(\xi) = 2\xi + a.$$

satisfy the necessary and sufficient condition of consistency if

$$\rho(1) = a + b + 1 = 0 \quad \text{and} \quad \rho'(1) = 2 + a = \sigma(1).$$

Thus, the method is consistent for $b = -1 - a$ and for any value of the parameter a.

The method is also stable if the roots of the characteristic equation

$$\rho(\xi) = \xi^2 + a\xi - 1 - a = 0$$

lie in the unit circle and those with modulus one are simple. Since

$$\xi_1 \xi_2 = -1 - a,$$

the method is stable if $|1 + a| \leq 1$, i.e.,

$$-2 < a \leq 0 \quad \text{and} \quad -1 \leq b < 1.$$

Let us note that the method is unstable for $a = -2$, since then the polynomial $\rho(\xi) = (\xi - 1)^2$ has double root $\xi_1 = \xi_2 = 1$ on the unit circle.

The order $p = 2$ if

$$C_0 = \alpha_0 + \alpha_1 + 1 = a + b + 1 = 0,$$

$$C_1 = (\alpha_1 + 2) - (\beta_0 + \beta_1 + \beta_2) = (a + 2) - (a + 2) = 0,$$

$$C_2 = \tfrac{1}{2}(\alpha_1 + 4) - (\beta_1 + 2\beta_2) = \tfrac{1}{2}(a + 4) - 2 = 0.$$

Hence, for $a = 0$ and $b = -1$, the method has order $p = 2$ and is consistent and stable.

2.2.6   Convergence

We consider convergence of the approximate solution y_n determined by a linear multistep method (2.3) to the theoretical solution $y(x)$ of the initial value problem (2.1), (2.2) when $h \to 0$, (i.e., $h = \frac{1}{N} \to 0$ when $N \to \infty$). Evidently, any linear $k-$ step method requires k starting values $y_0, y_1, \ldots, y_{k-1}$. In general, the starting values can be evaluated by another methods and therefore they are biased by some errors. However, these values should tend to the initial value $y_0 = y_a$ when $h \to 0$.

Let us assume that Cauchy-Picard's theorem is applicable to the initial value problem (2.1), (2.2) i.e., it has a unique continuously differentiable solution $y(x)$. Then, the following definition of convergence is used:

Definition 2.5 *A linear $k-$ step method is said to be convergent if there exists a limit*

$$\lim_{h \to 0, nh = x - a} y_n = y(x)$$

for all $x \in [a, b]$ and for all starting values

$$y_0 = y_a, \quad y_1 = \epsilon_1(h), \quad y_2 = \epsilon_2(h), \quad \ldots, y_{k-1} = \epsilon_{k-1}(h)$$

such that

$$\lim_{h \to 0} \epsilon_r(h) = y_a, \quad r = 1, 2, \ldots, k - 1.$$

The following theorem holds:

Theorem 2.2 *A consistent and stable method is convergent.*

The proof of this theorem is given in [15], p.244. We shall state and prove a relevant theorem for the two step methods.

$$y_{n+2} + \alpha_1 y_{n+1} + \alpha_0 y_n = h[\beta_2 f_{n+2} + \beta_1 f_{n+1} + \beta_0 f_n]. \tag{2.22}$$

As it is known, any consistent method of the form (2.22) can reach order $p \leq 4$ and the only method of order $p = 4$ is the Simpson's method. Let $y(x)$ be the $(p+1)$ times continuously differentiable solution of equation (2.1). Then, the following theorem holds:

Theorem 2.3 *If the linear two step method (2.22) of order p is consistent and stable and if $1 - h\beta_2 > 0$ then the method is convergent and the global error of the method $e_n = y(x_n) - y_n$, $n = 0, 1, ..., N$, satisfies the inequality*

$$|e_n| \leq \frac{6}{1 - \alpha_0}[|\delta| + |C_{p+1}|M^{(p+1)}h^p]e^{(x_n - a)K}, \quad n = 0, 1, ..., N,$$

where $\delta = y(x_1) - y_1$ is the error at the starting point x_1, $\beta = max\{|\beta_0|, |\beta_1|, |\beta_2|\}$,

$$K = \frac{6\beta}{(1 - \alpha_0)(1 - h\beta_2)}, \quad M^{(p+1)} = (b - a)|C_{p+1}| \sup |y^{(p+1)}(x)|.$$

Proof. Let us first note that $\alpha_0 \neq 1$. Indeed, because of consistency $\alpha_1 = -1 - \alpha_0$ and

$$\rho(\xi) = \xi^2 - (1 + \alpha_0)\xi + \alpha_0 = (\xi - 1)(\xi - \alpha_0).$$

By assumption, the method is stable, therefore $\xi_1 = 1$ must be a simple root, so that, the another root $\xi_2 = \alpha_0 \neq 1$. We shall now show the following:

1. (a) Let $\gamma_0, \gamma_1, ...$; be the coefficients determined by the identity

$$\frac{1}{\alpha_0 \xi^2 + \alpha_1 \xi + 1} = \gamma_0 + \gamma_1 \xi + \gamma_2 \xi^2 + \cdots \tag{2.23}$$

Then

$$\gamma_n = \frac{1 - \alpha_0^{n+1}}{1 - \alpha_0}, \tag{2.24}$$

and for $-1 \leq \alpha_0 < 1$,

$$0 < \gamma_n \leq \frac{2}{1 - \alpha_0}, \quad n = 0, 1, ...$$

(b) Let z_m be the solution of the difference equation

$$z_{m+2} + \alpha_1 z_{m+1} + \alpha_0 z_m = h[\beta_2 z_{m+2} + \beta_1 z_{m+1} + \beta_0 z_m] + \lambda_m, \quad (2.25)$$

for $m = 0, 1, ...$ and $z_0 = 0$, $z_1 = \delta$.
Then

$$|z_m| \le K_N e^{(m-1)hK}, \quad m = 0, 1, ...N,$$

where

$$K_N = \frac{6}{(1-\alpha_0)(1-h\beta_2)}[|\delta| + \lambda(N-1)], \quad \lambda = \max\{|\lambda_0|, |\lambda_1|, ..., |\lambda_m|\}.$$

Multiplying (2.23) by $\alpha_0 \xi^2 + \alpha_1 \xi + 1$ and comparing the coefficients at $1, \xi, \xi^2, ...$, we obtain the following difference equation:

$$\alpha_0 \gamma_n + \alpha_1 \gamma_{n+1} + \gamma_{n+2} = 0, \quad n = 0, 1, ... \quad (2.26)$$

$$\gamma_0 = 1, \quad \alpha_1 \gamma_0 + \gamma_1 = 0.$$

Solving this difference equation for $\alpha_1 = -1 - \alpha_0$, we find

$$\gamma_n = \frac{1 - \alpha_0^{n+1}}{1 - \alpha_0}, \quad \gamma_0 = 1, \quad \gamma_1 = 1 + \alpha_0.$$

By the stability condition $-1 \le \alpha_0 < 1$. Therefore

$$0 < \gamma_n \le \frac{2}{1 - \alpha_0}, \quad n = 0, 1, ...$$

Let us now multiply equation (2.25) by γ_l for $l = 0, 1, .., n-2$, $m = n - l - 2$ and add the resulting equations.
Then, from the left, we get

$$L_n = (z_n + \alpha_1 z_{n-1} + \alpha_0 z_{n-2})\gamma_0 + (z_{n-1} + \alpha_1 z_{n-2} + \alpha_0 z_{n-3})\gamma_1 + \cdots$$

$$+(z_2 + \alpha_1 z_1 + \alpha_0 z_0)\gamma_{n-2}.$$

Hence, by rearrangement of the terms

$$L_n = \gamma_0 z_n + (\alpha_1 \gamma_0 + \gamma_1)z_{n-1} + (\alpha_0 \gamma_0 + \alpha_1 \gamma_1 + \gamma_2)z_{n-2} + \cdots$$

$$+(\alpha_0 \gamma_{n-4} + \alpha_1 \gamma_{n-3} + \gamma_{n-2})z_2 + (\alpha_0 \gamma_{n-3} + \alpha_1 \gamma_{n-2})z_1 + \alpha_0 \gamma_{n-2} z_0,$$

By (2.24) and (2.26), the expression for L_n reduces to

$$L_n = z_n - \frac{1 - \alpha_0^n}{1 - \alpha_0}\delta = z_n - \gamma_{n-1}\delta. \quad (2.27)$$

From the right, we get

$$
\begin{aligned}
P_n = \; & h[(\beta_2 z_n + \beta_1 z_{n-1} + \beta_0 z_{n-2})\gamma_0 \\
& + (\beta_2 z_{n-1} + \beta_1 z_{n-2} + \beta_0 z_{n-3})\gamma_1 + \cdots \\
& + (\beta_2 z_2 + \beta_1 z_1 + \beta_0 z_0)\gamma_{n-2}] \\
& + \lambda_{n-2}\gamma_0 + \lambda_{n-3}\gamma_1 + \cdots + \lambda_0 \gamma_{n-2}.
\end{aligned}
\tag{2.28}
$$

Comparing (2.27) and (2.28), we obtain

$$
\begin{aligned}
(1 - h\beta_2)z_n = \; & \gamma_{n-1}\delta + h[(\beta_1 z_{n-1} + \beta_0 z_{n-2})\gamma_0 \\
& + (\beta_2 z_{n-1} + \beta_1 z_{n-2} + \beta_0 z_{n-3})\gamma_1 + \cdots \\
& + (\beta_2 z_2 + \beta_1 z_1 + \beta_0 z_0)\gamma_{n-2}] \\
& + \lambda_{n-2}\gamma_0 + \lambda_{n-3}\gamma_1 + \cdots + \lambda_0 \gamma_{n-2}.
\end{aligned}
\tag{2.29}
$$

Because $0 < 1 - h\beta_2 < 1$ and $-1 \leq \alpha_0 < 1$, it follows from (2.29) that

$$
|z_n| \leq \frac{6}{(1 - \alpha_0)(1 - h\beta_2)}[|\delta| + h\beta \sum_{j=0}^{n-1} |z_j| + \lambda(N - 1)].
\tag{2.30}
$$

If

$$
q = \frac{6}{(1 - \alpha_0)(1 - h\beta_2)}[|\delta| + \lambda(N - 1)] \quad \text{and} \quad r = \frac{6h\beta}{(1 - \alpha_0)(1 - h\beta_2)}
$$

then inequality (2.30) can be rewritten as follows:

$$
|z_n| \leq q + r \sum_{j=0}^{n-1} |z_j|.
\tag{2.31}
$$

Let us now show that

$$
|z_n| \leq q(1 + r)^{n-1}, \quad n = 1, 2, \ldots
$$

Indeed, we have $z_0 = 0$ and $|z_1| \leq q$.
By the principle of induction and from the assumption

$$
|z_{n-1}| \leq q + r \sum_{j=0}^{n-2} |z_j| \leq q(1 + r)^{n-2}
$$

via (2.31), we get

$$
|z_n| \leq q + r \sum_{j=0}^{n-2} |z_j| + r|z_{n-1}| \leq q(1 + r)^{n-2} + rq(1 + r)^{n-2} = q(1 + r)^{n-1}
$$

In terms of K, K_N, β and δ, this inequality takes the following form:

$$|z_m| \leq K_N (1 + hK)^{m-1}, \quad m = 0, 1, ..., N.$$

Hence, by

$$1 + hK \leq e^{hK},$$

we obtain the expected inequality

$$|z_m| \leq K_N e^{(m-1)hK}, \quad m = 0, 1, ..., N.$$

The theoretical solution $y(x)$ satisfies the equation

$$y(x_{n+2}) + \alpha_1 y(x_{n+1}) + \alpha_0 y(x_n) =$$

$$= h[\beta_2 f(x_{n+2}, y(x_{n+2})) + \beta_1 f(x_{n+1}, y(x_{n+1})) + \beta_0 f(x_n, y(x_n))] + \quad (2.32)$$

$$+ C_{p+1} y^{(p+1)}(\xi) h^{p+1},$$

for a certain $\xi \in (a, b)$, where C_{p+1} is the error constant.
Let $e_n = y(x_n) - y_n$, $n = 0, 1, ..., N$, be the error of the method. Subtracting equation (2.22) from equation (2.32) , we get

$$e_{n+2} + \alpha_1 e_{n+1} + \alpha_0 e_n = h[\beta_2 g_{n+2} e_{n+2} + \beta_1 g_{n+1} e_{n+1} + \beta_0 g_n e_n] + C_{p+1} y^{(p+1)}(\xi) h^{p+1},$$
$$(2.33)$$

where

$$g_n = \begin{cases} [f(x_n, y(x_n)) - f(x_n, y_n)] e_n^{-1} & if \quad e_n \neq 0, \\ 0 & if \quad e_n = 0. \end{cases}$$

In view of the Lipschitz condition

$$|g_n| \leq L, \quad n = 0, 1, ...$$

Now, in (b) let $m = n$, $z_n = e_n$ and $\lambda_n = C_{p+1} y^{(p+1)}(\xi) h^{p+1}$, we get

$$|e_n| \leq K_N e^{(n-1)hK}, \quad n = 0, 1, ..., N,$$

Because $\lambda = |C_{p+1}| M^{(p+1)} h^{p+1}$, therefore

$$(N - 1)\lambda \leq (b - a)|C_{p+1}| M^{(p+1)} h^p$$

and

$$K_N = \frac{6}{(1 - \alpha_0)(1 - h\beta_2)} [|\delta| + (b - a)|C_{p+1}| M^{(p+1)} h^p].$$

Hence, for $nh = x_n - a$, we obtain the desire estimate

$$|e_n| \leq \frac{6}{1 - \alpha_0} [|\delta| + |C_{p+1}| M^{(p+1)} h^p] e^{(x_n - a)K}, \quad n = 0, 1, ..., N.$$

2.3   Runge-Kutta Methods.

All Runge-Kutta methods of order $p \geq 2$ are nonlinear one step methods. They are nonlinear, since the function $f(x, y)$ is not involved in the Runge-Kutta formulas as a combination of its values, but it emerges as itself argument. However, the explicit Euler method is the one step linear method, and it is also the Runge-Kutta method of order $p = 1$. We shall consider Runge-Kutta explicit methods in the following general form:

$$y_{n+1} = y_n + h_n \Phi(x_n, y_n, h_n), \quad n = 0, 1, ..., N - 1, \qquad (2.34)$$

where N determines the number of mesh points

$$a = x_0 < x_1 < x_2 < \cdots < x_n < x_{n+1} < \cdots < x_N = b,$$

in the interval $[a, b]$ which are not necessarily uniformly distributed. However, they are distributed in such a way that

$$\lim sup_{N \to \infty} \frac{\max_{0 \leq n < N}(x_{n+1} - x_n)}{\min_{0 \leq n \leq N-1}(x_{n+1} - x_n)} = \sigma > 0.$$

The function $\Phi(x, y, h)$ which defines a Runge-Kutta method will be determined later on. At this point, we shall put on $\Phi(x, y, h)$ the following condition of consistency:

$$\Phi(x, y, 0) = f(x, y), \quad \text{for all} \quad a \leq x \leq b, \quad c \leq y \leq d. \qquad (2.35)$$

Let us consider the operator

$$S[y, h](x) = y(x + h) - y(x) - h\Phi(x, y(x), h),$$

associated with Runge-Kutta methods (2.34). This operator is used to define the order of a Runge-Kutta method according to the following definition:

Definition 2.6 *A Runge-Kutta method is said to have order p if p is the largest integer for which the following equality holds:*

$$S[y, h](x) = y(x + h) - y(x) - h\Phi(x, y(x), h) = O(h^{p+1}),$$

for all solutions $y(x)$ of the differential equation (2.1).

Here, symbol $O(h^{p+1})$ denotes the class of all functions which are decaying to zero at the rate h^{p+1} when $h \to 0$, i.e., if $u \in O(h^{p+1})$ then

$$\lim_{h \to 0} \sup \frac{u(h)}{h^{p+1}} = constant \neq 0.$$

It becomes clear that, a consistent Runge-Kutta method in the sense of condition (2.35) has an order of at least $p = 1$. Indeed, we have

$$S[y, h](x) = y(x + h) - y(x) - h\Phi(x, y, h) = hy'(x) - h\Phi(x, y, h) + O(h^2).$$

Hence, by the condition of consistency

$$S[y, h](x) = h[\Phi(x, y, 0) - \Phi(x, y, h)] + O(h^2) = O(h^2),$$

provided that $\Phi(x, y, h)$ is a differentiable function.

2.3.1 Formula for the function $\Phi(x, y, h)$ in the general form

Let us note that the initial value problem(2.1), (2.2) is equivalent with the following integral equation:

$$y(x) = y(a) + \int_a^x f(t, y(t))dt, \quad a \le x \le b.$$

Therefore

$$y(x_{n+1}) = y(x_n) + \int_{x_n}^{x_{n+1}} f(t, y(t))dt, \quad n = 0, 1, ... N - 1.$$

According to the general methods of numerical integration

$$\int_{x_n}^{x_{n+1}} f(t, y(t))dt = h \sum_{j=0}^{m} a_{mj} f(\xi_j, \eta_j) + R_m(h, f), \tag{2.36}$$

where

$R_m(h, f)$ *is the truncation error,*

$h = x_{n+1} - x_n,$

$\xi_j = x_n + \gamma_j h, \quad 0 \le \gamma_j \le 1, \quad j = 0, 1, ..., m,$

$$\tag{2.37}$$

$\eta_0 = y(x_n), \quad \eta_{j+1} = \eta_0 + h \sum_{i=0}^{j} a_{ji} f(\xi_i, \eta_i), \quad j = 0, 1, ... m - 1,$

m is a number of points in the subinterval $[x_n, x_{n+1}],$

determined by the method of integration (2.36).

Thus, the theoretical solution $y(x)$ satisfies the following equation:

$$y(x_{n+1}) = y(x_n) + h \sum_{j=0}^{m} a_{mj} f(\xi_j, \eta_j) + R_m(h, f). \tag{2.38}$$

By canceling the truncation error $R_m(h, f)$ in (2.38), we arrive at the general form of the explicit Runge-Kutta method

$$y_{n+1} = y_n + h \sum_{j=0}^{m} a_{mj} f(\xi_j, \nu_j), \quad n = 0, 1, ..., N - 1, \tag{2.39}$$

where

$$\Phi(x_n, y_n, h) = \sum_{j=0}^{m} a_{mj} f(\xi_j, \nu_j),$$

$$\xi_j = x_n + \gamma_j h, \quad j = 0, 1, ..., m; \quad h = x_{n+1} - x_n,$$

$$\nu_0 = y_n, \quad \nu_{j+1} = \nu_0 + h \sum_{i=0}^{j} a_{ji} f(\xi_i, \nu_i), \quad j = 0, 1, ..., m-1.$$

2.3.2 Derivation of Runge-Kutta methods

In order to obtain a Runge-Kutta method, we should determine the coefficients a_{mj}, a_{ji} and γ_i, $i = 0, 1, ..., j$; $j = 0, 1, ..., m$. We can find these coefficients assuming that the order of the method is equal to p. Then, the truncation error $R_m(h, f) = O(h^{p+1})$. On the other hand, by equality (2.38), we have

$$R_m(h, f) = y(x_{n+1}) - y(x_n) - h \sum_{j=0}^{m} a_{mj} f(\xi_j, \eta_j).$$

Hence, the method has order p if

$$R_m(h, f) = S[y, h](x_n) = O(h^{p+1}).$$

By Maclaurin's formula

$$R_m(h, f) = R_m(0, f) + R_m'(0, f)h + R_m''(0, f)\frac{h^2}{2!} + \cdots$$
$$+ R_m^{(p)}(0, f)\frac{h^p}{p!} + R_m^{(p+1)}(\zeta, f)\frac{h^{p+1}}{(p+1)!}.$$

Clearly

$$R_m(h, f) = O(h^{p+1}),$$

if

$$R_m(0, f) = 0, \ R_m'(0, f) = 0, \dots, R_m^{(p)}(0, f) = 0, \text{ and } R_m^{(p+1)}(0, f) \neq 0. \quad (2.40)$$

Thus, a Runge-Kutta method has order p if its coefficients a_{mj}, a_{ji}, γ_i, $i = 0, 1, ...j$, $j = 0, 1, ..., m-1$, satisfy conditions (2.40).
Below, we shall derive some of Runge-Kutta methods.

2.3.3   Runge-Kutta method of order one

Let us put $p = 1$, $m = 0$ in (2.39) and (2.40). Then, we obtain the following subclass of Runge-Kutta methods of order one:

$$y_{n+1} = y_n + ha_{00}f(\xi_0, \nu_0), \quad n = 0, 1, ..., N - 1,$$

$$\xi_0 = x_n, \quad \nu_0 = y_n.$$

In this case, there is only one free parameter a_{00}, so that the truncation error

$$R_0(h, f) = y(x + h) - y(x) - ha_{00}f(x, y(x)).$$

By conditions (2.40)

$$R_0(0, f) = 0 \quad \text{and} \quad R_0'(0, f) = 0 \quad \text{but} \quad R_0''(0, f) \neq 0.$$

Since

$$R_0'(h, f) = y'(x + h) - a_{00}f(x, y(x)),$$

therefore

$$R_0(0, f) = 0, \quad R_0'(0, f) = y'(x) - a_{00}f(x, y(x)) = (1 - a_{00}f(x, y(x))) = 0.$$

Hence $a_{00} = 1$, and the explicit Runge-Kutta method of order $p = 1$ is:

$$y_{n+1} = y_n + hf_n, \quad y_0 = y_a, \quad n = 0, 1, ..., N - 1.$$

The explicit Euler method has the local truncation error

$$R_0(h, f) = \frac{1}{2}h^2 y''(\zeta), \quad \zeta \in (a, b).$$

2.3.4   Runge-Kutta method of order two

Let us put $p = 2$, $m = 1$ in (2.39) and (2.40). Then, we obtain the following subclass of Runge-Kutta methods:

$$y_{n+1} = y_n + ha_{10}f(\xi_0, \nu_0) + ha_{11}f(\xi_1, \nu_1), \quad n = 0, 1, ..., N - 1,$$

$$\xi_0 = x_n, \quad \xi_1 = x_n + \gamma_1 h, \quad \nu_0 = y_n, \quad \nu_1 = y_n + ha_{00}f(\xi_0, \nu_0).$$

This subclass of Runge-Kutta methods contains four free parameters

$$a_{00}, \ a_{10}, \ a_{11} \quad and \ \gamma_1$$

and its truncation error is:

$$R_1(h, f) = y(x + h) - y(x) - ha_{10}f(\xi_0, \nu_0) - ha_{11}f(\xi_1, \nu_1).$$

In order to determine coefficients a_{00}, a_{10}, a_{11} and γ, we find

$$R_1'(h, f) = y'(x+h) - a_{10}f(\xi_0, \nu_0) - a_{11}f(\xi_1, \nu_1) - ha_{11}[\frac{\partial f}{\partial x}\gamma_1 + f\frac{\partial f}{\partial y}a_{00}],$$

$$R_1''(h, f) = y''(x+h) - 2a_{11}[\frac{\partial f}{\partial x}\gamma_1 + f\frac{\partial f}{\partial y}a_{00}]-$$

$$ha_{11}[\frac{\partial^2 f}{\partial x \partial x}\gamma_1^2 + 2f\frac{\partial^2 f}{\partial x \partial y}\gamma_1 a_{00} + f^2\frac{\partial^2 f}{\partial y^2}a_{00}^2].$$

Hence, by condition (2.40), when $h = 0$ and $y''(x) = \frac{\partial f}{\partial x} + f\frac{\partial f}{\partial y}$, we obtain

$$R_1(0, f) = 0,$$

$$R_1'(0, f) = (1 - a_{10} - a_{11})f = 0,$$

$$R_1''(0, f) = (1 - 2\gamma_1 a_{11})\frac{\partial f}{\partial x} + (1 - 2a_{00}a_{11})f\frac{\partial f}{\partial y} = 0.$$

The above conditions hold for all functions f, if

$$
\begin{aligned}
a_{10} + a_{11} &= 1 \\
2\gamma_1 a_{11} &= 1 \\
2a_{00}a_{11} &= 1
\end{aligned}
\tag{2.41}
$$

Let us note that the system of three algebraic equations (2.41) with four unknown parameters a_{00}, a_{10}, a_{11}, and γ_1 possesses an infinite number of solutions. This means that, there are an infinite number of Runge-Kutta methods of order $p = 2$. Some of these method are given below.

2.3.5   Mid-point Euler method

Clearly, the values

$$a_{10} = 0, \quad a_{11} = 1, \quad a_{00} = \frac{1}{2}, \quad \gamma_1 = \frac{1}{2}$$

satisfy the equations (2.41). Therefore, the Runge-Kutta method of order $p = 2$ is:

$$y_{n+1} = y_n + hf(x_n + \frac{1}{2}h, y_n + \frac{1}{2}hf(x_n, y_n)), \quad y_0 = y_a, \quad n = 0, 1, ..., N - 1.$$

Now, let us choose

$$a_{10} = \frac{1}{2}, \quad a_{11} = \frac{1}{2}, \quad a_{00} = 1, \quad \gamma_1 = 1.$$

Then, we get another Runge-Kutta method of order $p = 2$

$$y_{n+1} = y_n + \frac{h}{2}[f(x_n, y_n) + f(x_n + h, y_n + hf(x_n, y_n))], \quad y_0 = y_a, \quad n = 0, 1, ..., N-1.$$

Example 2.3 *Consider the following orbit system of equations (cf.[4])*

$$\frac{du_1}{dx} = -\alpha u_1 - \beta u_2$$

$$\frac{du_2}{dx} = \beta u_1 - \alpha u_2$$

with the initial value conditions $u_1(0) = 1$, $u_2(0) = 0$.

We shall solve this initial value problem using `Mathematica` module `oneStepKR2` which is an implementation of the Mid-Point method of order $p = 2$.

Program 2.2 *Mathematica module that implements Runge-Kutta method of order two*

```
oneStepRK2[rhSides_, xuvariables_, xuvariables0_, h_] :=
  Module[{x,vars,x0,vars0, k1, k2 },
    {x,vars}={First[xuvariables],Rest[xuvariables]};
    {x0,vars0}={First[xuvariables0],Rest[xuvariables0]};
    k1 = h N[ rhSides /.x->x0/. Thread[vars -> vars0] ];
    k2 = h N[ rhSides /. x->x0+h/2/.Thread[vars -> vars0 + k1/2]];
    Prepend[vars0 + k2, x0+h]
  ]
```

To invoke the module `oneStepRK2`, we use the following definition:

```
rungeKutta[rhSides_List, xuvariables_List,
           xuvariables0_List, h_, n_] :=
  NestList[oneStepRK2[rhSides, xuvariables, #, N[h]]&,
    N[xuvariables0], n ]
```

We input data

```
alpha=0.5; beta=2;
system1={-alpha u1 -beta u2,beta u1 -alpha u2};
```

Then, we activate the above `NestList` by the command

```
kuttasol=rungeKutta[system1,{x,u1,u2},{0,1,0},0.1,60];
```

In order to obtain the table of both the exact and approximate solutions, we execute the following `Mathematica` instructions:

```
kuttaexactTable=N[Table[
   {kuttasol[[i+1,1]],
    kuttasol[[i+1,2]],
    N[E^(-alpha i h) Cos[beta i h],4],
    kuttasol[[i+1,3]],
    N[E^(-alpha i h) Sin[beta i h],4]},
                      {i,0,50}]];
   h=0.1;   TableForm[N[kuttaexactTable],
            TableHeadings->{{ },
            {"x","u1-RKutta", "u1", "u2-RKutta", "u2"}}]
```

To show graphics of both exact and approximate soluions, we execute following the instructions:

```
   rungePlot=ListPlot[Map[Rest[#]&, kuttasol],
            PlotJoined->True]

   exactPlot=ParametricPlot[{E^(-al x) Cos[be x],
            E^(-al x) Sin[be x]},{x,0,6}]

   r1=Show[exactPlot, Ticks->{{-0.5,0,0.5,1},{-0.3,0.3,0.6}}]

   r2=Show[rungePlot,Ticks->{{-0.5,0,0.5,1},{-0.3,0.3,0.6}}]

   Show[GraphicsArray[{r1,r2}]]
```

Note that the idea of Runge-Kutta method is simple and theoretically one can easily get any Runge-Kutta algorithm. However, it is extremely difficult technically, to obtain a Runge-Kutta method of order $p \geq 10$. The number of arithmetic operations involved in a Runge-Kutta algorithm increases exponentially with respect to its order p. Higher order Runge-Kutta methods are expensive in terms of arithmetic operations and may produce results strongly biased by round-off errors. In practice, Runge-Kutta algorithms of order four are considered to be the most efficient. [1] However, theoretically, all explicit consistent Runge-Kutta methods are convergent.

2.3.6   Runge-Kutta method of order four

The Runge-Kutta methods of order four are most often used and they are within the subclass of the following form:

$$y_{n+1} = y_n + h \sum_{j=0}^{3} a_{3j} f(\xi_j, \nu_j), \quad n = 0, 1, ..., N - 1,$$

---

[1]When 8-digit arithmetic is used

where $\xi_0 = x_n$, $\xi_1 = x_n + \gamma_1 h$, $\xi_2 = x_n + \gamma_2 h$, $\xi_3 = x_n + \gamma_3 h$,

$$\begin{aligned}
\nu_0 &= y_n, \\
\nu_1 &= \nu_0 + h a_{00} f(x_n, y_n), \\
\nu_2 &= \nu_0 + h[a_{10} f(x_n, y_n) + a_{11} f(\xi_1, \nu_1)], \\
\nu_3 &= \nu_0 + h[a_{20} f(x_n, y_n) + a_{21} f(\xi_1, \nu_1) + a_{22} f(\xi_2, \nu_2)].
\end{aligned}$$

This subclass of Runge-Kutta methods contains thirteen free parameters γ_1, γ_2, γ_3, $a_{3,j}$, $j = 0, 1, 2, 3$, and a_{10}, a_{20}, a_{11}, a_{21}, a_{22}. These parameters are determined by the order $p = 4$ of the method for which the local truncation error

$$R_3(h, f) = y(x + h) - y(x) - h \sum_{j=0}^{3} a_{3j} f(\xi_j, \eta_j).$$

As we know, the method has order $p = 4$ if the local truncation error satisfies the conditions:

$$R_3(0, f) = 0, \quad R_3'(0, f) = 0, \quad R_3''(0, f) = 0, \quad R_3'''(0, f) = 0, \quad R_3^{(4)}(0, f) = 0.$$

One can check that the above conditions hold for the parameters which satisfy the following system of nonlinear equations:

$$\gamma_1 = a_{00}$$

$$\gamma_2 = a_{10} + a_{11}$$

$$\gamma_3 = a_{20} + a_{21} + a_{22}$$

$$a_{30} + a_{31} + a_{32} + a_{33} = 1$$

$$a_{31}\gamma_1 + a_{32}\gamma_2 + a_{33}\gamma_3 = \frac{1}{2}$$

$$a_{31}\gamma_1^2 + a_{32}\gamma_2^2 + a_{33}\gamma_3^2 = \frac{1}{3}$$

$$a_{31}\gamma_1^3 + a_{32}\gamma_2^3 + a_{33}\gamma_3^3 = \frac{1}{4}$$

$$a_{32}a_{11}\gamma_1 + a_{33}a_{21}\gamma_1 + a_{33}a_{22}\gamma_2 = \frac{1}{6}$$

$$a_{32}a_{11}\gamma_1\gamma_2 + a_{33}a_{21}\gamma_1\gamma_3 + a_{33}a_{22}\gamma_2\gamma_3 = \frac{1}{8}$$

$$a_{32}a_{11}\gamma_1^2 + a_{33}a_{21}\gamma_1^2 + a_{33}a_{22}\gamma_2^2 = \frac{1}{12}$$

$$a_{33}a_{22}a_{11}\gamma_1 = \frac{1}{24}$$

Every solution of the above system of equations determines a Runge-Kutta method of order $p = 4$. Let us consider the following solution

$$\gamma_1 = \frac{1}{2}, \quad \gamma_2 = \frac{1}{2}, \quad \gamma_3 = 1,$$

$$a_{00} = \frac{1}{2}, \quad a_{10} = 0, \quad a_{11} = \frac{1}{2},$$

$$a_{20} = 0, \quad a_{21} = 0, \quad a_{22} = 1,$$

$$a_{30} = \frac{1}{6}, \quad a_{31} = \frac{1}{3}, \quad a_{32} = \frac{1}{3}, \quad a_{33} = \frac{1}{6}.$$

Then, we obtain the following Runge-Kutta method of order $p = 4$

$$y_{n+1} = y_n + \frac{1}{6}(k_1 + 2k_2 + 2k_3 + k_4), \quad n = 0, 1, ..., N - 1, \tag{2.42}$$

where

$$k_1 = hf(x_n, y_n)$$

$$k_2 = hf(x_n + \frac{h}{2}, y_n + \frac{k_1}{2})$$

$$k_3 = hf(x_n + \frac{h}{2}, y_n + \frac{k_2}{2})$$

$$k_4 = hf(x_n + h, y_n + k_3).$$

To obtain another variant of Runge-Kutta method of order $p = 4$, let us choose the following solution

$$\gamma_1 = \frac{1}{4}, \quad \gamma_2 = \frac{1}{2}, \quad \gamma_3 = 1,$$

$$a_{00} = \frac{1}{4}, \quad a_{10} = 0, \quad a_{11} = \frac{1}{2},$$

$$a_{20} = 1, \quad a_{21} = -2, \quad a_{22} = 2,$$

$$a_{30} = \frac{1}{6}, \quad a_{31} = 0, \quad a_{32} = \frac{2}{3}, \quad a_{33} = \frac{1}{6}.$$

Then, the Runge-Kutta method that corresponds to the above coefficients is:

$$y_{n+1} = y_n + \frac{1}{6}(k_1 + 4k_3 + k_4), \quad n = 0, 1, ..., N - 1,$$

where

$$k_1 = hf(x_n, y_n)$$

$$k_2 = hf(x_n + \frac{h}{4}, y_n + \frac{k_1}{4})$$

$$k_3 = hf(x_n + \frac{h}{2}, y_n + \frac{k_2}{2})$$

$$k_4 = hf(x_n + h, y_n + k_1 - 2k_2 + 2k_3).$$

Below, we give the `Mathematica` module `oneStepRK4` which solves a system of ordinary differential equations by the Runge-Kutta method of order $p = 4$.

Program 2.3 *Mathematica module that implements Runge-Kutta method of order four*

```
oneStepRK4[rhSides_, xuvariables_, xuvariables0_, h_] :=
    [{x,uvars,x0,uvars0, k1, k2, k3, k4 },
    {x,uvars}={First[xuvariables],Rest[xuvariables]};
    {x0,uvars0}={First[xuvariables0],Rest[xuvariables0]};
    k1 = h N[ rhSides /.x->x0/. Thread[uvars -> uvars0] ];
    k2 = h N[ rhSides /. x->x0+h/2/.Thread[uvars -> uvars0 + k1/2] ];
    k3 = h N[ rhSides /. x->x0+h/2/.Thread[uvars -> uvars0 + k2/2] ];
    k4 = h N[ rhSides /.x->x0+h/. Thread[uvars -> uvars0 + k3] ];
    Prepend[uvars0 + (k1 + 2 k2 + 2 k3 + k4)/6, x0+h]
    ]
```

In order to invoke the module `oneStepRK4`, we can use the following `Mathematica` definition:

```
rungeKutta[rhSides_, xuvariables_, xuvariables0_List, h_, n_] :=
    NestList[
            oneStepRK4[rhSides, xuvariables, #, N[h]]&,
            N[xuvariables0], n ]
```

To obtain a numerical solution, we input data as in the module `oneStepRK2`.

2.3.7 Convergence of Runge-Kutta methods

We shall now state and prove the following theorem on convergence:

Theorem 2.4 *A Runge-Kutta method of order $p \geq 1$ is convergent and the global error of the method*

$$e_n(h) = y(x_n) - y_n, \quad n = 0, 1, ..., N,$$

satisfies the following inequality:

$$|e_n(h)| \leq [e^{(x_n-a)K\sigma} - 1]M \, h^p, \quad n = 0, 1, ..., N,$$

where K and M are constants independent of the step size h.

Proof. Let $y(x)$ be the theoretical solution of the initial value problem (2.1), (2.2) and let y_n, $n = 0, 1, ..., N$ be the approximate solution determined by a Runge-Kutta method (2.34) of the order $p \geq 1$. Then, the following equations hold:

$$y(x_{n+1}) = y(x_n) + h \sum_{j=0}^{m} a_{mj} f(\xi_j, \eta_j) + R_m(h, f), \quad n = 0, 1, ..., N-1,$$
$$h = x_{n+1} - x_n, \quad \xi_j = x_n + \gamma_j h,$$

$$\eta_0 = y(x_n), \quad \eta_{j+1} = \nu_0 + h \sum_{i=0}^{j} a_{ji} f(\xi_i, \eta_i),$$

(2.43)

and

$$y_{n+1} = y_n + h \sum_{j=0}^{m} a_{mj} f(\xi_j, \nu_j), \quad n = 0, 1, ..., N-1,$$

$$\nu_0 = y_n, \quad \nu_{j+1} = \nu_0 + h \sum_{i=0}^{j} a_{ji} f(\xi_i, \nu_i).$$

(2.44)

Let us subtract equation (2.44) from equation (2.43). Then, we obtain

$$e_{n+1} = e_n + h \sum_{j=0}^{m} [f(\xi_j, \eta_j) - f(\xi_j, \nu_j)] + R_m(h, f).$$

(2.45)

Hence, by the Lipschitz condition and the triangle inequality

$$|e_{n+1}| \leq |e_n| + hL \sum_{j=0}^{m} |a_{mj}||\eta_j - \nu_j| + |R_m(h, f)|, \quad n = 0, 1, ..., N-1. \quad (2.46)$$

Let

$$z_j = h \sum_{i=0}^{j} |\eta_i - \nu_i|, \quad j = 0, 1, ..., m.$$

Below, we shall estimate z_m. Let us therefore note that

$$z_j - z_{j-1} = h|\eta_j - \nu_j|$$

$$= h|(y(x_n) - y_n) + h \sum_{i=0}^{j-1} a_{j-1i}[f(\xi_i, \eta_i) - f(\xi_i, \nu_i)]|$$

$$\leq h[|e_n| + hAL \sum_{i=0}^{j-1} |\eta_i - \nu_i|] = h[|e_n| + ALz_{j-1}],$$

where $A = \max\limits_{0 \leq i \leq j, \, 0 \leq j \leq m} |a_{ji}|$ and L is the Lipschitz constant.
Hence, we obtain the inequality

$$z_j \leq h|e_n| + (1 + hAL)z_{j-1}, \quad j = 0, 1, ..., m.$$

(2.47)

Denoting by
$$r = h|e_n|, \qquad q = 1 + hAL,$$
we write inequality (2.47) in terms of r and q
$$z_j \leq r + qz_{j-1}, \quad j = 0, 1, ..., m. \tag{2.48}$$

Let us now show that
$$z_m \leq r\frac{1 - q^{m+1}}{1 - q}.$$

Indeed, $z_0 = h|\eta_0 - \nu_0| = h|e_n| = r$. By the principle of induction, we assume
$$z_{m-1} \leq r\frac{1 - q^m}{1 - q}$$

Hence, via (2.48), we obtain
$$z_m \leq r + qz_{m-1} \leq r + qr\frac{1 - q^m}{1 - q} = r\frac{1 - q^{m+1}}{1 - q},$$

and
$$z_m \leq h|e_n|\frac{(1 + hAL)^{m+1} - 1}{hAL}. \tag{2.49}$$

Clearly, there exists a constant K, for fixed m, such that
$$\frac{(1 + hAL)^{m+1} - 1}{hAL} =$$
$$\binom{m+1}{1} + \binom{m+1}{2}hAL + \cdots + \binom{m+1}{m+1}(hAL)^m \leq K.$$

Hence, by (2.49), we have the inequality
$$z_m \leq hK|e_n|. \tag{2.50}$$

Now, combining inequalities (2.46) and (2.50), we obtain
$$|e_{n+1}| \leq |e_n| + hK|e_n| + |R_m(h, f)|, \quad n = 0, 1, ..., N - 1. \tag{2.51}$$

By the assumption, Runge-Kutta method has order p. Therefore, there exists a constant K_0 independent of h such that
$$|R_m(h, f)| \leq K_0 h^{p+1}.$$

Hence, by (2.51)
$$|e_{n+1}| \leq (1 + hK)|e_n| + K_0 h^{p+1}, \quad n = 0, 1, ..., N - 1. \tag{2.52}$$

Now, let us denote by
$$r = K_0 h^{p+1}, \qquad q = 1 + hK.$$

Then, inequality (2.52), takes the following form:

$$|e_{n+1}| \le r + q|e_n|, \qquad n = 0, 1, ..., N - 1. \tag{2.53}$$

Let us now show that

$$|e_n| \le \frac{1 - q^n}{1 - q}.$$

Indeed, we have $e_0 = 0$ and $|e_1| \le r + q|e_0| = r$. By the principle of induction, we assume that

$$|e_{n-1}| \le r\frac{1 - q^{n-1}}{1 - q}$$

Hence, via (2.53), we obtain

$$|e_n| \le r + q|e_{n-1}| \le r + qr\frac{1 - q^{n-1}}{1 - q} = r\frac{1 - q^n}{1 - q},$$

and

$$|e_{n+1}| \le = K_0 h^{n+1} \frac{(1 + hK)^{n+1} - 1}{hK}. \tag{2.54}$$

Because

$$h = x_{n+1} - x_n \le \sigma \, min(x_{n+1} - x_n) \le \sigma\frac{x_n - a}{n},$$

therefore

$$(1 + hK)^{n+1} \le e^{(x_n - a)K\sigma}, \tag{2.55}$$

By (2.54), we obtain the required inequality

$$|e_{n+1}| \le [e^{(x_n - a)K\sigma} - 1]M \, h^p, \qquad n = 0, 1, ..., N - 1,$$

where the constants K and $M = \dfrac{K_0}{K}$ are independent of h.

2.4   Boundary Value Problem

Let us consider the following system of differential equations (cf. [30])

$$-\frac{d^2u}{dx^2} = g(x, u), \qquad a \le x \le b, \tag{2.56}$$

with the boundary conditions

$$u(a) = u_0, \qquad u(b) = u_{n+1},$$

where $u(x) = (u^{(1)}(x), u^{(2)}(x), ..., u^{(p)}(x))$ is the unknown function, and $g(x, u) = (g^{(1)}(x, u), g^{(2)}(x, u), ..., g^{(p)}(x, u))$, is given.

This boundary value problem has a unique regular solution if the function

$g(x, u)$ is continuous and possesses partial derivatives, such that, the Jacobian matrix $J(g^{(1)}, g^{(2)}, ..., g^{(p)})$ is non-positive definite, that is

$$(J(g^{(1)}, g^{(2)}, ..., g^{(p)})\lambda, \lambda) \leq 0,$$

for all $\lambda = (\lambda_1, \lambda_2, ..., \lambda_p)$, $x \in (a, b)$ and $-\infty < u^{(q)} < \infty$, $q = 1, 2, ..., p$.

Finite difference scheme. We shall approximate the system of equations (2.56) by the following finite difference scheme with the global error $O(h^4)$ (cf. [5, 36])

$$-L_h v_i \quad = \quad g(x_i, v_i), \qquad i = 1, 2, ..., n.$$

$$v_0 = u_0, \qquad v_{n+1} = u_{n+1}, \tag{2.57}$$

where $v_i = v(x_i)$, $x_i = a + ih$, $i = 0, 1, ..., n + 1$, $h = \dfrac{b - a}{n + 1}$, and the finite difference operator L_h is given by formula (see (3.7), (3.9)

$$L_h v_i \equiv \begin{cases} \dfrac{v_{i-1} - 2v_i + v_{i+1}}{h^2}, & i = 1, n, \\[3mm] \dfrac{-v_{i-2} + 16v_{i-1} - 30v_i + 16v_{i+1} - v_{i+2}}{12h^2}, & i = 2, 3, ..., n - 1. \end{cases}$$

Let us write the finite difference scheme (2.57) in matrix form as

$$\mathbf{A}v = f(v), \tag{2.58}$$

where $v = (v^1, v^2, ..., v^p)^T$, with $v^{(q)} = (v_1^{(q)}, v_2^{(q)}, ..., v_n^{(q)})$, $f = (f^{(1)}, f^{(2)}, ..., f^{(p)})$, with $f^{(q)} = (f_1^{(q)}, f_2^{(q)}, ..., f_n^{(q)})$, $q = 1, 2, ..., p$, and

$$f_i^{(q)} = 12h^2 \begin{cases} g^{(q)}(x_i, v_i) + \dfrac{u_0}{h^2}, & if \quad i = 1, \\[3mm] g^{(q)}(x_i, v_i) - \dfrac{u_0}{12h^2}, & if \quad i = 2, \\[3mm] g^{(q)}(x_i, v_i), & if \quad i = 3, 4, ..., n - 2, \\[3mm] g^{(q)}(x_i, v_i) - \dfrac{u_{n+1}}{12h^2}, & if \quad i = n - 1, \\[3mm] g^{(q)}(x_i, v_i) + \dfrac{u_{n+1}}{h^2}, & if \quad i = n. \end{cases}$$

The matrix $\mathbf{A}$ is a block diagonal matrix of the form $\mathbf{A} = diagonal(M)$, where M is the pentadiagonal matrix

$$M = \begin{bmatrix} 24 & -12 & 0 & 0 & 0 & 0 & \cdots & 0 & 0 \\ -16 & 30 & -16 & 1 & 0 & 0 & \cdots & 0 & 0 \\ 1 & -16 & 30 & -16 & 1 & 0 & \cdots & 0 & 0 \\ 0 & 1 & -16 & 30 & -16 & 1 & \cdots & 0 & 0 \\ \vdots & \vdots & \vdots & \vdots & \vdots & \vdots & \ddots & \vdots & \vdots \\ 0 & 0 & 0 & 0 & 0 & 0 & \cdots & 30 & -16 \\ 0 & 0 & 0 & 0 & 0 & 0 & \cdots & -12 & 24 \end{bmatrix}_{n \times n}$$

2.4.1   Algorithm

The matrix equation (2.58) splits into p decoupled systems of equations of the form

$$Mv^{(q)} = f^{(q)}, \quad q = 1, 2, ..., p. \tag{2.59}$$

Applying Gauss elimination to the specific pentadiagonal system of linear eaquations (2.59), for fixed q, we obtain the following algorithm:

1.

$$\alpha_1 = \frac{12}{24}, \quad \alpha_2 = \frac{16}{22},$$

$$\beta_1 = 0, \quad \beta_2 = \frac{1}{22},$$

$$d_1 = 24, \quad d_2 = 22.$$

2. For $i = 3, 4, .., n-1$, evaluate

$$c_i = 16 + \alpha_{i-2},$$

$$d_i = 30 - \beta_{i-2} - \alpha_{i-1}c_i,$$

$$\beta_i = \frac{1}{d_i},$$

$$\alpha_i = \frac{16 - \beta_{i-1}c_i}{d_i}.$$

3.

$$b_1 = \frac{f_2}{24}, \qquad\qquad b_2 = \frac{f_2 + 16b_1}{d_2},$$

$$b_i = \frac{f_i - b_{i-2} + b_{i-1}c_i}{d_i}, \quad i = 3, 4, ..., n-1,$$

$$b_n = \frac{f_n + 12b_{n-1}}{24 - 12\alpha_{n-1}}.$$

4.

$$x_n = b_n,$$

$$x_{n-1} = b_{n-1} + \alpha_{n-1}x_n,$$

$$x_i = b_i + \alpha_i x_{i+1} - \beta_i x_{i+2}, \quad i = n-2, n-3, ..., 2,$$

$$x_1 = b_1 + \alpha_1 x_2.$$

2.4.2 Implicit Iterations

In order to solve the system of algebraic equations (2.58) or equivalent systems in(2.59), we apply the following implicit iterations:

$$\mathbf{A}v^{(m+1)} = f(v^{(m)}), \quad m = 0, 1, ..., s, \tag{2.60}$$

where the starting value of v, $v^{(0)}$, is given.

Since $\mathbf{A}$ is a block diagonal matrix, the above system splits into p decoupled systems of equations

$$Mv^{(q,m+1)} = f^{(q)}(v^{(q,m)}), \quad q = 1, 2, ..., p. \tag{2.61}$$

Because the pentagonal matrix M is monotonic, the iterative method is convergent, (cf. [39, 43]).

To solve the system of equations (2.59), we apply the following steps:

> For $m = 1, 2, ..., s$,
>
> 1. set $v = v^{(m-1)}$,
>
> 2. compute $\mathbf{f} = (f^{(1)}(v), f^{(2)}(v), ..., f^{(p)}(v))$,
>
> 3. solve (2.61) by the algorithm given in the previous section.

2.4.3 Mathematica Module

Several systems of differential equations of the form (2.56) have been solved using the algorithm with *Mathematica*. The module `solveBVP` takes three optional parameters `bound, startv` and `iters`, in addition to the parameters g, p, a, b, and n. The parameter `bound` specifies the boundary conditions $\{u_0, u_{n+1}\}$ as a $p \times 2$ array, `startv`$=v^{(0)}$ is a $p \times n$ array, and `iters`$=s$. The default values of the optional parameters correspond to the homogeneous boundary conditions.

Program 2.4 *Mathematica module that solves the boundary value problem*

---

bound— $p \times 2$ array of zeros corresponding to the homogeneous boundary conditions,
startv — a $p \times n$ array of zeros,
iters $= 2n$.

```
Options[solveBVP] = {bound  -> homogeneous,
                     startv -> vzero,
                     iters  -> twon};

solveBVP[g_, p_, a_, b_, n_, opts___]:= Module[
        { h, x, boundary, v0, iterNumber, beta, gamma,
```

```
      solution, bcorrections, useboundary,
      solve3, solve5, oneStep  },

h = (b-a)/(n+1);
x = Table[ N[a + i h], {i, 1, n}];
homogeneous = Table[0, {p}, {2}];
vzero = Table[0,{p},{n}];
twon = 2n;

boundary=bound/.{opts}/.Options[solveBVP];
v0=startv/.{opts}/.Options[solveBVP];
iterNumber = iters/.{opts}/.Options[solveBVP];

beta={0};
Do[AppendTo[beta, 1/(14-Last[beta]) ], {3}];
Do[AppendTo[beta, N[7 - 4 Sqrt[3]] ], {n-3}];
gamma=Table[i/(i+1), {i, 1, n}];
bcorrections=Table[
    {{12 *h^2 *boundary[[q,1]], - h^2*boundary[[q,1]]},
     {-h^2*boundary[[q,2]], 12*h^2*boundary[[q,2]]}}/h^2,
    {q,1,p}];

useboundary[f_, bcorr_]:=
    Join[Take[f,2] + bcorr[[1]],
        Take[f, {3,-3}], Take[f,-2] + bcorr[[2]]];

solve3[f_, d11_, alpha_]:= Module[{ f1, sol },
    f1[1]=f[[1]]/d11;
    f1[i_]:=f1[i]=(f[[i]]+f1[i-1])*alpha[[i]];
    sol[n]=If[d11==12, f[[n]]/12, f1[n]];
    sol[i_]:=sol[i]=f1[i]+alpha[[i]]*sol[i+1];
    Table[sol[i],{i,1,n}]];
solve5[f_]:=With[
    {w= solve3[f, 12, beta]},
    solve3[w, 2, gamma] ];

oneStep[v_]:=Module[ {ff, bff},
    ff=12*h^2* N[Transpose[MapThread[g,{x,Transpose[v]}]]];
    bff=MapThread[useboundary, {ff,  bcorrections}];
    Map[solve5, bff] ];

solution = Nest[oneStep, v0, iterNumber];
PrependTo[solution, x];
```

```
With[{initiala=
        Prepend[Table[boundary[[q,1]], {q,1,p}],a],
      endb=
        Prepend[Table[boundary[[q,2]], {q,1,p}],b]},
      Append[Prepend[Transpose[solution],initiala],endb]]
]
```

Numerical Examples. We present two examples of the boundary value problem (2.56) which are solved by the Module `solveBVP`.

Example 2.4 *Find the solution of the equation that represents rotation of a heavy string (cf. [11])*

$$\frac{d^2u}{dx^2} + \frac{u}{4\sqrt{x^2+u^2}} = 0, \quad 0 < x < 1, \tag{2.62}$$

with the boundary value conditions $u(0) = 0$, *and* $u(1) = 1$.

We consider the partition of the interval $[a,b] = [0,1]$ into $n = 9$ subintervals by $n+1$ points. So that, $p = 1$, $a = 0$, $b = 1$, $n = 9$, and the list of the boundary value conditions $\{u_0, u_{n+1}\} = \{\{0,1\}\}$.

To solve equation (2.62), we define the function

```
g1[x_,u_]:={u[[1]]/(4 Sqrt[x^2+u[[1]]^2])}
```

and call the module

```
solveBVP[g1,1,0,1,9,bound->{{0,1}}]//TableForm
```

In the table, we present numerical results.

x	v(x)
0	0
0.1	0.108172
0.2	0.214509
0.3	0.319017
0.4	0.421704
0.5	0.522577
0.6	0.621644
0.7	0.718912
0.8	0.814389
0.9	0.908082
1.	1.

One can easily find the interpolating polynomial

$$P(x) = 1.09093x - 0.092147x^2 + 0.001197x^3 - 0.0000155x^4 + 0.000131x^5 -$$

$$0.000239x^6 + 0.000293x^7 - 0.0002295x^8 + 0.0001044x^9 - 0.0000209x^{10}$$

through the data points given in the above table using *Mathematica*. This polynomial is the approximate solution that satisfies the equation (2.62) in the interval $(0, 1)$, with the residual error $\approx 10^{-5}$. Because the finite difference scheme (2.57) is $O(h^4)$ convergent, the global error $u(x) - P(x) \approx O(h^4)$, when $0 < x < 1$.

Example 2.5 *Find the solution of the system of two nonlinear equations that represent the equilibrium state of a rotating rod (cf. [11]).*

$$
\begin{aligned}
\frac{d^2 u^{(1)}}{dx^2} &= \frac{1}{2} \sin u^{(2)}(x), \\
\frac{d^2 u^{(2)}}{dx^2} &= \frac{1}{2} u^{(1)}(x) \cos u^{(2)}(x), \quad 0 < x < 1,
\end{aligned}
\tag{2.63}
$$

with the nonhomogeneous boundary value conditions

$$
u^{(1)}(0) = 0, \; u^{(1)}(1) = 1, \quad u^{(2)}(0) = 0, \; u^{(2)}(1) = 1.
$$

In this example, we define the function

```
g2[x_, u_] :={-Sin[u[[2]]]/2,-u[[1]]*Cos[u[[2]]]/2}
```

and execute the command

```
solveBVP[g2,2,0,1,9,bound->{{0,1},{0,1}}]//TableForm
```

The numerical results are given in the following table

x	v1(x)	v2(x)
0	0	0
0.1	0.0924969	0.0932606
0.2	0.185459	0.186982
0.3	0.279351	0.281612
0.4	0.374632	0.377582
0.5	0.471755	0.47529
0.6	0.571165	0.575089
0.7	0.673294	0.67728
0.8	0.778553	0.782086
0.9	0.887335	0.889646
1	1	1

The interpolating polynomials

$$
\begin{aligned}
P(x) = \; & 0.924189x + 0.0000862252x^2 + 0.0768919x^3 + 0.0041257x^4 - 0.0157484x^5 + \\
& 0.0322844x^6 - 0.0476032x^7 + 0.0430081x^8 - 0.0222084x^9 + 0.00497499x^{10}
\end{aligned}
$$

and

$$Q(x) = \; 0.931819x + 0.000375188x^2 + 0.0739647x^3 + 0.0156642x^4 - 0.0604811x^5 +$$

$$0.115398x^6 - 0.166608x^7 + 0.149199x^8 - 0.0762911x^9 + 0.01696x^{10}$$

correspond to data $x, v1, v2$ in the table. One can check, with use of *Mathematica*, that these polynomials satisfy equilibrium equations in the interval $(0, 1)$, with the residual error $\approx 10^{-4}$. So that, the global errors $u^{(1)}(x) - P(x) = O(h^4)$, and $u^{(2)}(x) - Q(x) = O(h^4)$.

2.5   Exercises

Question 2.1 .

1. (a) *Find the order and determine the local truncation error of the method*

$$y_{n+4} - \frac{8}{19}(y_{n+3} - y_{n+1}) - y_n = \frac{6h}{19}(f_{n+4} + 4f_{n+3} + 4f_{n+1} + f_n),$$

$$n = 0, 1, \ldots, N - 4.$$

   (b) *Investigate convergence of the method.*

Question 2.2 *Solve the initial value problem*

$$y'(x) = -\lambda y(x), \quad y(0) = 1, \quad 0 \leq x \leq 1,$$

by the following methods:

1. (a) *the explicit Euler method,*
   (b) *the implicit Euler method,*
   (c) *the trapezian method,*
   (d) *the mid-point method,*
   (e) *the Simpson's method,*

for $\lambda = 1, \, 6, \, 12$ and for $h = 0.05, \, 0.1$.
Compare the theoretical solution $y(x) = exp(-\lambda x)$ with the approximate solutions.

Question 2.3 *Consider the following six step method:*

$$\alpha_0 y_n + \alpha_1 y_{n+1} + \alpha_2 y_{n+2} + \alpha_3 y_{n+3} + \alpha_4 y_{n+4} + \alpha_5 y_{n+5} + y_{n+6} =$$

$$h[\beta_0 f_n + \beta_1 f_{n+1} + \beta_2 f_{n+2} + \beta_3 f_{n+3} + \beta_4 f_{n+4} + \beta_5 f_{n+5} + \beta_6 f_{n+6},$$

$$n = 0, 1, ..., N - 6.$$

1. (a) Use the `Mathematica` module `multistepmethod` to find the method of the highest order within the class of the six step methods.

   (b) Show that the method you found in (a) is consistent and stable.

   (c) Determine the error constant and the local truncation error.

Question 2.4 Show that the linear operator

$$P[y](x) = \sum_{j=0}^{k} [\alpha_j y(x + jh) - h\beta_j y'(x + jh)]$$

has order p if and only if

$$P[x^r] = 0 \quad \text{for} \quad r = 0, 1, ..., p,$$

and $P[x^{p+1}] \neq 0$.

Question 2.5 Find the range of the parameter t for which the two-step method

$$y_{n+2} - (3+t)y_{n+1} + (2+t)y_n = \frac{h}{2}[(1-t)f_{n+1} - (3+t)f_n], \quad n = 0, 1, \ldots, N-2,$$

is stable.
Show that the method has order $p = 3$ if $t = -7$. Also, show that in this the method is unstable.

Question 2.6 Show that the method

$$y_{n+3} + (t - 3)(y_{n+2} - y_{n+1}) - y_n = \frac{th}{2}[f_{n+2} + f_n], \quad n = 0, 1, \ldots, N-3.$$

satisfies the necessary and sufficient conditions of consistency for all values of the parameter t. Find the range of the parameter t for which the method is stable. Determine the value of the parameter t for which the method has the highest possible order.

Question 2.7 Find the solution y_n, $n = 0, 1, ...,$ of the difference equation

$$y_{n+2} - y_{n+1} = \frac{h}{12}[4f_{n+2} + 8f_{n+1} - f_n], \quad n = 0, 1, ...,$$

when $f(x, y) = 1$.
Show that the approximate solution y_n, $n = 0, 1, ...,$ is not convergent to the theoretical solution $y(x) = x$ of the initial problem

$$y'(x) = 1, \quad y(0) = 0.$$

Explain why y_n does not converge to $y(x)$.

Question 2.8 *Show that Runge-Kutta method*

$$y_{n+1} = y_n + \frac{h}{4}[f(x_n, y_n) + 3f(x_n + \frac{2h}{3}, y_n + \frac{2h}{3}f(x_n, y_n))], \quad n = 0, 1, ..., N-1.$$

has order $p = 2$. Determine the local truncation error of the method.

Question 2.9 *Derive a Runge-Kutta method of order $p = 3$.*

Question 2.10 *Following the proof of the theorem on convergence of Runge-Kutta methods show that the mid-point Euler method is quadratically convergent.*

Question 2.11 *Solve the initial value problem*

$$y'(x) = x^2, \quad y(0) = 0, \quad 0 \le x \le 1,$$

by a Runge-Kutta method of the order $p = 2$, when $h = 0.2$. Show that any Runge-Kutta method of order $p \ge 3$ produces the exact solution $y(x)$.

Question 2.12 *Consider the following initial value problem:*

$$\frac{dy}{dx} = x^p, \quad y(a) = y_0, \quad a \le x \le b.$$

where $p \ge 0$ is an integer.

1. (a) *Show that every Runge-Kutta method of order greater than p produces exact solution of the initial value problem.*

   (b) *Solve the initial value problem for $p = 3$, that is*

   $$y'(x) = x^3, \quad y(0) = 0, \quad 0 \le x \le 1,$$

   by the Runge-Kutta method of the order $p = 4$,

   $$y_{n+1} = y_n + \frac{1}{6}(k_1 + 2k_2 + 2k_3 + k_4),$$

   for $y(0) = 1$, $h = 0.2$, and $n = 0, 1, 2, 3, 4$.
   Here, we have

   $$k_1 = h\, f(x_n, y_n), \qquad\qquad k_2 = h\, f(x_n + \frac{h}{2}, y_n + \frac{k1}{2}),$$

   $$k_3 = h\, f(x_n + \frac{h}{2}, y_n + \frac{k_2}{2}), \quad k_4 = h\, f(x_n + h, y_n + k_3),$$

   for $f(x, y) = x^3$.

(c) Compare the solution y_n, $n = 0, 1, 2, 3, 4, 5$ with the exact solution $y(x) = 1 + \dfrac{x^3}{3}$ at the knots $x_0 = 0$, $x_1 = 0.2$, $x_2 = 0.4$, $x_3 = 0.6$, $x_4 = 0.8$ and $x_5 = 1$. From your calculations, conclude whether the statement, that is established in (a), holds for this equation and the method.

Question 2.13 *Consider the following boundary value problem:*

$$-\frac{d^2 u(x)}{dx^2} = e^{-u(x)}, \quad 0 \le x \le 1,$$

$$u(0) = 0, \quad u(1) = 0.$$

1. (a) Use `Mathematica` *module* `solveBVP` *to solve the boundary value problem.*

   (b) *Give the approximate solution in the form of an interpolating polynomial.*

   (c) *Tabulate the residual error using step size* $h = 0.05$.

Chapter 3

Finite Difference Method

Tadeusz Styš
University of Warsaw

Abstract: Two techniques of analysis of convergence and the global error estimates have been developed. The first technique is based on the discrete maximum principle to prove uniform convergence of finite difference methods for elliptic and parabolic equations. The second technique draws on spectral analysis and deals with average convergence in the discrete Hilbert's space H. The chapter ends with a set of questions.

Chapter 3

Finite Difference Method

3.1   Introduction

Finite difference method has a long history. In 1768, [7], Euler used the simplest finite difference procedure to replace the derivative in the equation $\dfrac{du}{dx} = f(x, u)$ by the finite difference $\dfrac{u_{i+1} - u_i}{\Delta x}$ at point $x = x_i$. Further applications of the method have been established by Runge in 1908 [23] and by Richardson in 1910 [21]. They used finite differences to approximate solution to the Poisson's equation $u_{xx} + u_{yy} = f$. Later on, fundamental results were published by Gerschgorin [11], Mikeladze [18], Collatz [6], Forsythe and Wasow [9]. However, the greatest development of the method came together with computers. At this point we could give a long list of authors that made substantial contributions to both the theoretical and practical aspects of the method . Although, the finite difference method is well established it still inspires new results and is successfully used to solve problems from various areas of mathematical sciences.

3.2   Finite difference approximation of derivatives

Finite difference approximation of derivatives is associated with a partition of the interval $[a, b]$ by points x_i, $i = 0, 1, ..., N$. Such a partition, we shall call a network and denote by Ω_h, so that

$$\Omega_h : \quad a = x_0 < x_1 < ... < x_i < x_{i+1} < ... < x_N = b.$$

$a = x_0 \qquad\qquad x_{i-2} \quad\ x_{i-1} \quad\ x_i \quad\ x_{i+1} \quad\ x_{i+2} \qquad\qquad x_N = b$

Fig. 3.1. Network Ω_h

To approximate first and second derivatives of a differentiable function $u(x)$, we shall use the partition of $[a, b]$ by points $x_i = a+ih$, $i = 0, 1, ..., N$, $Nh = b-a$, uniformly distributed in $[a, b]$. Then, with $u_i = u(x_i)$, by Taylor's formula

$$
\begin{aligned}
u_{i-1} &= u_i - u_i' \frac{h}{1!} + u_i''(\eta_i) \frac{h^2}{2!}, && \eta_i \in (x_{i-1}, x_i), \\
u_{i+1} &= u_i + u_i' \frac{h}{1!} + u_i''(\xi_i) \frac{h^2}{2!}, && \xi_i \in (x_i, x_{i+1}).
\end{aligned}
\tag{3.1}
$$

Hence, we obtain the following finite difference approximations of the first derivative:

The forward finite difference approximation:

$$
\Delta u_i = \frac{u_{i+1} - u_i}{h} = u_i' + u''(\xi_i) \frac{h}{2},
\tag{3.2}
$$

for a certain $\xi_i \in (x_i, x_{i+1})$, where the local truncation error

$$
\psi_i(h) = \frac{h}{2} u''(\xi_i).
$$

The backward finite difference approximation:

$$
\nabla u_i = \frac{u_i - u_{i-1}}{h} = u_i' - u''(\eta_i) \frac{h}{2},
\tag{3.3}
$$

for a certain $\eta_i \in (x_{i-1}, x_i)$, where the local truncation error

$$
\psi_i(h) = -\frac{h}{2} u''(\eta_i).
$$

Let us note that the above forward and backward finite difference approximations hold for any function $u(x)$ which is twice continuously differentiable in the interval $[a, b]$. However, if $u(x)$ has third continuous derivative in $[a, b]$, then

$$
\begin{aligned}
u_{i-1} &= u_i - u_i' \frac{h}{1!} + u_i'' \frac{h^2}{2!} - u_i''' (\eta_i) \frac{h^3}{3!}, && \eta_i \in (x_{i-1}, x_i), \\
u_{i+1} &= u_i + u_i' \frac{h}{1!} + u_i'' \frac{h^2}{2!} + u_i''' (\xi_i) \frac{h^3}{3!}, && \xi_i \in (x_i, x_{i+1}),
\end{aligned}
\tag{3.4}
$$

and using the intermediate value theorem, we get
the central finite difference approximation:

$$
\delta u_i = \frac{u_{i+1} - u_{i-1}}{2h} = u_i' + \frac{h^2}{6} u'''(\zeta_i),
\tag{3.5}
$$

for a certain $\zeta_i \in (x_{i-1}, x_{i+1})$, where the local truncation error

$$
\psi_i(h) = \frac{h^2}{6} u'''(\zeta_i).
$$

Assuming now that $u(x)$ is four times continuously differentiable, we can use Taylor's formula up to five terms, i.e.

$$u_{i-1} = u_i - u_i'\frac{h}{1!} + u_i''\frac{h^2}{2!} - u_i'''\frac{h^3}{3!} + u^{(4)}(\eta_i)\frac{h^4}{4!}, \qquad \eta_i \in (x_{i-1}, x_i),$$
$$u_{i+1} = u_i + u_i'\frac{h}{1!} + u_i''\frac{h^2}{2!} + u_i'''\frac{h^3}{3!} + u^{(4)}(\xi_i)\frac{h^4}{4!}, \qquad \xi_i \in (x_i, x_{i+1}). \tag{3.6}$$

Hence, we get the
Tri-point finite difference approximation of the second derivative.

$$\Lambda u_i \equiv \frac{u_{i-1} - 2u_i + u_{i+1}}{h^2} = u_i'' + \frac{h^2}{12}u^{(4)}(\zeta_i), \tag{3.7}$$

for a certain $\zeta_i \in (x_{i-1}, x_{i+1})$, where the local truncation error

$$\psi_i(h) = \frac{h^2}{12}u^{(4)}(\zeta_i), \qquad \zeta_i \in (x_{i-1}, x_{i+1}).$$

We can also approximate the second derivative $u''(x_i)$ by a five-point $O(h^4)$ accurate finite difference scheme. For such a scheme, we assume that the function $u(x)$ is six times continuously differentiable in interval $[a, b]$. Then, by Taylor's formula

$$u_{i\pm1} = u_i \pm \frac{h}{1!}u_i' + \frac{h^2}{2!}u_i'' \pm \frac{h^3}{3!}u_i''' + \frac{h^4}{4!}u_i^{(4)} \pm \frac{h^5}{5!}u_i^{(5)} + \frac{h^6}{6!}u^{(6)}(\xi_i),$$

with $\zeta_i \in (x_{i-1}, x_{i+1})$,

$$u_{i\pm2} = u_i \pm \frac{2h}{1!}u_i' + \frac{2^2h^2}{2!}u_i'' \pm \frac{2^3h^3}{3!}u_i''' + \frac{2^4h^4}{4!}u_i^{(4)} \pm \frac{2^5h^5}{5!}u_i^{(5)} + \frac{2^6h^6}{6!}u^{(6)}(\eta_i),$$

with $\zeta_i \in (x_{i-2}, x_{i+2})$,

$$\tag{3.8}$$

Hence, by the intermediate value theorem

$$L_h u_i \equiv \frac{-u_{i-2} + 16u_{i-1} - 30u_i + 16u_{i+1} - u_{i+2}}{12h^2} = u_i'' + \frac{h^4}{90}u^{(6)}(\zeta_i), \tag{3.9}$$

where the local truncation error

$$\psi_i(h) = \frac{h^4}{90}u^{(6)}(\zeta_i), \qquad \zeta_i \in (x_{i-2}, x_{i+2}).$$

Example 3.1 *Write down an $O(h^2)$ finite difference scheme for the following boundary value problem:*

$$-\frac{d^2u(x)}{dx^2} + c(x)u = f(x), \quad 0 \le x \le l,$$
$$u(0) = \alpha, \qquad u(l) = \beta, \tag{3.10}$$

where $c(x) \ge 0$, $f(x)$ are given continuous functions in interval $[0, l]$ and α and β are known boundary values of $u(x)$. Estimate the local truncation error.

Solution. In order to approximate the boundary value problem (3.10), we define the network

$$\Omega_h = \{x_i = ih \ : \quad i = 1, ..., N-1, \quad Nh = l\ \}$$

with the boundary points $x_0 = 0$ and $x_N = l$.

Assuming that $u(x)$ is a solution of the boundary value problem (3.10) four times continuously differentiable, we can use the finite difference operator Λ to approximate derivative $\dfrac{d^2 u(x_i)}{dx^2}$.

Clearly, the theoretical solution $u(x)$ satisfies the difference equation

$$-\Lambda u_i + c_i u_i = f_i + \frac{h^2}{12} \frac{d^4 u(\zeta_i)}{dx^4}, \qquad i = 1, 2, ..., N-1.$$

Canceling the local truncation error

$$\psi_i(h) = \frac{h^2}{12} \frac{d^4 u(\zeta_i)}{dx^4},$$

we arrive at the following $O(h^2)$ finite difference scheme

$$-\Lambda v_i + c_i v_i = f_i, \qquad i = 1, 2, ..., N-1,$$

$$v_0 = \alpha, \qquad v_N = \beta.$$

Obviously, the local truncation error $\psi_i(h)$ satisfies the inequality

$$|\psi_i(h)| \le \frac{M^{(4)}}{12} h^2, \qquad i = 1, 2, ..., N-1,$$

where the constant

$$M^{(4)} = \sup_{0 \le x \le l} |\frac{d^4 u(x)}{dx^4}|.$$

3.3   Tri-points $O(h^4)$ Finite Difference Approximation

In special cases, it is possible to use tri-point scheme to obtain a higher order finite difference approximation. For example, let us assume that $f(x)$ is a four times continuously differentiable function. Consequently, the solution of the differential equation

$$-\frac{d^2 u(x)}{dx^2} = f(x), \qquad 0 \le x \le l,$$

is six times continuously differentiable and $u(x)$ satisfies the difference equation

$$-\Lambda u_i = f_i + \frac{h^2}{12} \frac{d^4 u_i}{dx^4} + \frac{h^4}{360} \frac{d^6 u(\zeta_i)}{dx^6}, \qquad \zeta_i \in (x_{i-1}, x_{i+1}), \qquad i = 1, 2, ..., N-1.$$

Since

$$\frac{d^4 u_i}{dx^4} = \frac{d^2 f_i}{dx^2},$$

we have

$$-\Lambda u_i = f_i + \frac{h^2}{12}\frac{d^2 f_i}{dx^2} + \frac{h^4}{360}\frac{d^6 u(\zeta_i)}{dx^6}, \qquad \zeta_i \in (x_{i-1}, x_{i+1}), \qquad i = 1, 2, ..., N-1.$$

Hence, we obtain $O(h^4)$ finite difference scheme

$$-\Lambda v_i = f_i + \frac{h^2}{12}\frac{d^2 f_i}{dx^2}, \qquad i = 1, 2, ..., N-1,$$

$$v_0 = \alpha, \qquad v_n = \beta, \tag{3.11}$$

with the local truncation error

$$\psi_i(h) = \frac{h^4}{360}\frac{d^6 u(\zeta_i)}{dx^6}, \qquad \zeta_i \in (x_{i-1}, x_{i+1}).$$

Substituting into (3.11)

$$\frac{d^2 f_i}{dx^2} = \Lambda f_i + \frac{h^2}{12}\frac{d^4 f(\xi_i)}{dx^4}, \quad \xi_i \in (x_{i-1}, x_{i+1}),$$

we shall arrive at the Numerov scheme

$$-\Lambda v_i = \frac{1}{12}(f_{i-1} + 10f_i + f_{i+1}), \quad i = 1, 2, ..., N-1,$$

$$v_0 = \alpha, \quad v_N = \beta,$$

with the local truncation error

$$\psi_i(h) = \frac{h^4}{144}\left[\frac{d^4 f(\xi_i)}{dx^4} + \frac{2}{5}\frac{d^6 u(\zeta_i)}{dx^6}\right].$$

In general, the local truncation error ψ of a difference scheme determines the global error of the method $v - u$, provided that such a scheme is stable. There are a few techniques to estimate the error of the method. We shall provide some estimates based on Hilbert's space properties and on the discrete maximum principle.

3.4   Hilbert Space of Discrete Functions.

Let us consider the set H_h of all discrete functions given on the closed network:

$$\overline{\Omega}_h = \{x_i = ih : \quad i = 0, 1, ..., N, \quad Nh = l\}.$$

A discrete function $v \in H_h$ defined on the network can be identified with a vector

$$v = (v_0, v_1, ..., v_N) \in R^{N+1}.$$

For such discrete functions $v, w \in H_h$ the following inner product:

$$(v, w) = h \sum_{i=0}^{N} v_i w_i.$$

is well defined.

The pair consisting of: the set H_h and the inner product $(.,.)$ is called *Hilbert's space of discrete functions*. One can check that $dim(H_h) = N + 1$, i.e.,there are exactly $N + 1$ linearly independent discrete functions in H_h which create a basis for the space H_h. The inner product $(.,.)$ generates the following norm:

$$\| v \| = \sqrt{(v, v)}.$$

Along with the Hilbert space H_h, we shall consider its subspace H_h^0 of all discrete functions which vanish at the ends of interval $[0, l]$, so that $v \in H_h^0$ if $v = (0, v_1, v_2, ..., v_{N-1}, 0)$.

Then, the inner product in the Hilbert space H_h^0 is given by the sum

$$(v, w) = h \sum_{i=1}^{N-1} v_i w_i \quad \text{for any} \quad v, w \in H_h^0.$$

This inner product generates the norm

$$\| v \| = \sqrt{(v, v)} = \sqrt{h \sum_{i=1}^{N-1} v_i^2}.$$

3.5 Eigenvalues and Eigenfunctions of the Difference Operator Λ

Let us state a few properties of the operator

$$\Lambda v_i \equiv \frac{v_{i-1} - 2v_i + v_{i+1}}{h^2}, \quad i = 1, 2, ..., N - 1,$$

which is defined on $v \in H_h^0$. We shall first find the eigenvalues and eigenfunctions of $-\Lambda$ solving the following eigenvalue problem:

Find all numbers λ and all non $-$ zero discrete functions $v \in H_h^0$

corresponding to λ such that

$$-\Lambda v_i = \lambda v_i, \quad i = 1, 2, ..., N - 1,$$

$$v_0 = 0, \quad v_N = 0.$$

$$(3.12)$$

Solution. We can rewrite equation (3.12) as follows:

$$(v_{i-1} + v_{i+1}) - 2(1 - \frac{\lambda h^2}{2})v_i = 0, \quad i = 1, 2, ..., N - 1,$$

$$v_0 = 0, \quad v_N = 0.$$

$$(3.13)$$

This equation has non-trivial solutions if the characteristic equation

$$\xi^2 - 2(1 - \frac{\lambda h^2}{2})\xi + 1 = 0$$

does not have real roots, i.e., when $0 < \lambda h^2 < 4$. Then, the non-trivial solutions have the following form:

$$v_i = \sin \alpha i h, \quad i = 0, 1, ..., N,$$

where α is a parameter.

Let us find α for which the equation (3.13) holds.

Obviously, α should satisfy the difference equation

$$\sin(i - 1)\alpha h + \sin(i + 1)\alpha h - 2(1 - \frac{\lambda h^2}{2}) \sin i\alpha h = 0, \quad i = 1, 2, ..., N - 1.$$

Hence

$$2(\cos \alpha h - 1 + \frac{\lambda h^2}{2}) \sin i\alpha h = 0, \quad i = 1, 2, ..., N - 1,$$

for

$$\lambda = \frac{2}{h^2}(1 - \cos \alpha h) = \frac{4}{h^2} \sin^2 \frac{\alpha h}{2}.$$

By the boundary condition

$$v_N = \sin \alpha N h = \sin \alpha l = 0,$$

implying that

$$\alpha_k = \frac{k\pi}{l}, \quad k = 1, 2, ..., N - 1.$$

Thus, we get the following eigenvalues:

$$\lambda_k = \frac{4}{h^2} \sin^2 \frac{k\pi h}{2l}, \quad k = 1, 2, ..., N - 1,$$

and eigenfunctions:

$$v_i^{(k)} = \sin \frac{k\pi i h}{l}, \quad k = 1, 2, ..., N - 1.$$

3.5.1   The Orthogonal Set of Eigenfunctions

The eigenfunctions

$$v_i^{(k)} = \sin \frac{k\pi i h}{l}, \qquad k = 1, 2, ..., N-1$$

are orthogonal in Hilbert's space H_h^0, i.e., the inner product

$$(v^{(k)}, v^{(m)}) = \begin{cases} 0 & if \ k \neq m, \\[2mm] \dfrac{l}{2} & if \ k = m. \end{cases}$$

Indeed, Λ is a symmetric operator in space H_h^0, since for any $v, w \in H_h^0$

$$\begin{aligned}
(\Lambda v, w) &= h \sum_{i=1}^{N-1} \Lambda v_i \, w_i \\
&= \frac{1}{h} \sum_{i=1}^{N-1} (v_{i-1} w_i - 2 v_i w_i + v_{i+1} w_i) \\
&= \frac{1}{h} \sum_{i=2}^{N} v_i w_{i-1} - \frac{2}{h} \sum_{i=1}^{N-1} v_i w_i + \frac{1}{h} \sum_{i=0}^{N-2} v_i w_{i+1} \\
&= \frac{1}{h} \sum_{i=1}^{N-1} v_i w_{i-1} - \frac{2}{h} \sum_{i=1}^{N-1} v_i w_i + \frac{1}{h} \sum_{i=1}^{N-1} v_i w_{i+1} \\
&= \frac{1}{h} \sum_{i=1}^{N-1} v_i (w_{i-1} - 2 w_i + w_{i+1}) \\
&= h \sum_{i=1}^{N-1} v_i \Lambda w_i = (v, \Lambda w).
\end{aligned}$$

We note that

$$\begin{aligned}
(\Lambda v^{(k)}, v^{(m)}) - (v^{(k)}, \Lambda v^{(m)}) &= -\lambda_k (v^{(k)}, v^{(m)}) + \lambda_m (v^{(k)}, v^{(m)}) \\
&= (\lambda_m - \lambda_k)(v^{(k)}, v^{(m)}) = 0.
\end{aligned}$$

Hence, for $k \neq m$, we have $\lambda_k \neq \lambda_m$ and $(v^{(k)}, v^{(m)}) = 0$.
But for $k = m$, we get

$$(v^k, v^k) = \| v^{(k)} \|^2 = h \sum_{j=1}^{N-1} v_j^{(k)} v_j^{(k)} = h \sum_{j=1}^{N-1} \sin^2 \frac{k\pi j h}{l}.$$

Using the identity

$$\sin^2 \frac{k\pi j h}{l} = \frac{1}{2}\Big(1 - \cos \frac{2k\pi j h}{l}\Big),$$

we can evaluate the norm $||v^{(k)}||$,

$$
\begin{aligned}
\| v^{(k)} \|^2 &= \frac{h}{2} \sum_{j=1}^{N-1} [1 - \cos \frac{2k\pi jh}{l}] \\
&= \frac{(N-1)h}{2} - \frac{h}{2} \sum_{j=1}^{N-1} Re \, \exp(i\frac{2k\pi jh}{l}) \\
&= \frac{l}{2} - \frac{h}{2} - \frac{h}{2}[-1 + Re \sum_{j=0}^{N-1} \exp(i\frac{2k\pi jh}{l})] \\
&= \frac{l}{2} + \frac{h}{2} Re \frac{1 - \exp(i2k\pi)}{1 - \exp(i2k\pi h)} = \frac{l}{2}.
\end{aligned}
$$

Thus, the norm of each eigenfunction

$$
\| v^{(k)} \| = \sqrt{\frac{l}{2}}, \quad k = 1, 2, ..., N-1.
$$

Therefore, the eigenfunctions

$$
v_i^{(k)} = \sqrt{\frac{2}{l}} \sin \frac{k\pi ih}{l}, \quad k = 1, 2, ..., N-1; \quad i = 0, 1, ..., N,
$$

are *orthonormal* in the Hilbert space H_h^0, i.e.,

$$
(v^{(k)}, v^{(m)}) = \begin{cases} 0 & if \quad k \neq m, \\ 1 & if \quad k = m. \end{cases} \tag{3.14}
$$

Let us note that the operator $-\Lambda$ has all its eigenvalues bounded below by the positive constant $\frac{8}{l^2}$, that is

$$
\lambda_k = \frac{4}{h^2} \sin^2 \frac{k\pi h}{2l} \geq \frac{8}{l^2}, \quad k = 1, 2, ..., N-1.
$$

Indeed, the sequence of angles $\frac{k\pi h}{2l}$, $k = 1, 2, ..., N-1$, is increasing to the angle $\frac{\pi}{2}$. Therefore, the sequence of eigenvalues is also increasing, so that

$$
\lambda_{N-1} > \lambda_{N-2} > \cdots > \lambda_2 > \lambda_1.
$$

Now, to estimate λ_1, we observe that

$$
\lambda_1 = \frac{4}{h^2} \sin^2 \frac{\pi h}{2l} = \frac{4}{h^2} (\frac{\pi h}{2l})^2 (\frac{\sin \frac{\pi h}{2l}}{\frac{\pi h}{2l}})^2 \tag{3.15}
$$

where the angle $0 < \dfrac{\pi h}{2l} \le \dfrac{pi}{4}$, provided that $h \le \dfrac{l}{2}$.

Thus, the function

$$f(x) = \frac{\sin x}{x}$$

is decreasing for $0 < x \le \frac{\pi}{4}$. Then, $f(x) \ge f(\frac{\pi}{4})$ and by (3.15)

$$\lambda_1 \ge \frac{8}{l^2}.$$

3.5.2   Fourier Series of a Discrete Function

The eigenfunctions $v^{(k)}$, $k = 1, 2, ..., N - 1$;   create a basis for the space H_h^0. Therefore, every discrete function $v \in H_h^0$ can be written as a linear combination:

$$v_i = \sum_{k=1}^{N-1} c_k v_i^{(k)}, \quad i = 0, 1, ..., N,$$

that is, as sum of a Fourier series with the Fourier coefficients are given by

$$c_k = (v, v^{(k)}), \quad k = 1, 2, ..., N - 1.$$

The following lemma holds:

Lemma 3.1 *(cf. [25]) Every discrete function $v \in H_h^0$ satisfies the inequality*

$$\frac{8}{l^2} \parallel v \parallel^2 \le (-\Lambda v, v) \le \frac{4}{h^2} \parallel v \parallel^2 \tag{3.16}$$

Proof of lemma. The eigenfunctions $v^{(k)}$, $k = 1, 2, ..., N - 1$; create a basis for the space H_h^0. Then, we can expand a function $v \in H_h^0$ in the Fourier series

$$v_i = \sum_{k=1}^{N-1} c_k v_i^{(k)}, \quad i = 0, 1, ..., N,$$

with the coefficients

$$c_k = (v, v^{(k)}), \quad k = 1, 2, ..., N - 1.$$

Clearly

$$(-\Lambda v, v) = \sum_{j,k=1}^{N-1} c_j c_k (-\Lambda v^{(k)}, v^{(j)}) = \sum_{j,k=1}^{N-1} c_k c_j \lambda_k (v^{(k)}, v^{(j)}).$$

Hence, by the orthogonality condition (3.14)

$$(-\Lambda v, v) = \sum_{k=1}^{N-1} \lambda_k c_k^2.$$

On the other hand

$$\sum_{k=1}^{N-1} c_k^2 = \sum_{k=1}^{N-1} c_k(v, v^{(k)}) = (v, \sum_{k=1}^{N-1} c_k v^{(k)}) = (v, v) = \| v \|^2 .$$

Thus

$$\lambda_1 \| v \|^2 = \lambda_1 \sum_{k=1}^{N-1} c_k^2 \leq (-\Lambda v, v) \leq \lambda_{N-1} \sum_{k=1}^{N-1} c_k^2 = \lambda_{N-1} \| v \|^2 .$$

Since $\lambda_1 \geq \dfrac{8}{l^2}$ and $\lambda_{N-1} < \dfrac{4}{h^2}$, we obtain the required inequality

$$\frac{8}{l^2} \| v \|^2 \leq (-\Lambda v, v) \leq \frac{4}{h^2} \| v \|^2 .$$

3.6   Positive Definite Operators

Let us now introduce the notion of *a positive definite operator in space H_h^0*.

Definition 3.1 *A linear operator* $L_h \; : \; H_h^0 \; \rightarrow \; H_h^0$ *is said to be positive definite in the space H_h^0 if and only if the following conditions hold:*

1. *(a) L_h is a symmetric operator, i.e. $(L_h v, w) = (v, L_h w)$ for any $v, w \in H_h^0$.*

    (b) There exists a positive constant $\gamma > 0$ such that

    $$(L_h v, v) \geq \gamma(v, v) \quad for \; any \; v \in H_h^0$$

In an infinite dimensional Hilbert's space H, we assume additionally that the operator L has its domain $D(L)$ dense in the space H.

Let us note that the positive definite operator $-\Lambda$ has the coherent matrix A given by the formula

$$-\Lambda v_i \rightarrow \frac{1}{h^2} \begin{bmatrix} 2 & -1 & 0 & \cdots & 0 & 0 \\ -1 & 2 & -1 & \cdots & 0 & 0 \\ \cdots & \cdots & \cdots & \cdots & \cdots & \cdots \\ 0 & 0 & 0 & \cdots & 2 & -1 \\ 0 & 0 & 0 & \cdots & -1 & 2 \end{bmatrix} \begin{bmatrix} v_1 \\ v_2 \\ \vdots \\ v_{N-2} \\ v_{N-1} \end{bmatrix} = Av.$$

for any $v \in H_h^0$.

Therefore, A is a positive definite and monotone matrix.

Example 3.2 *Let us recall the boundary value problem (3.10)*

$$-\frac{d^2u(x)}{dx^2} + c(x)u = f(x), \qquad 0 \le x \le l, \tag{3.17}$$

$$u(0) = \alpha, \qquad u(l) = \beta.$$

where $c(x) \ge 0$ and $f(x)$ are given continuous functions in the interval $[0, l]$, α and β are given boundary values of $u(x)$.

1. (a) *Approximate the boundary value problem (3.17) by an $O(h^2)$ finite difference scheme. Determine the local truncation error.*

   (b) *Estimate the error of the method in the norm of H_h^0.*

   (c) *Write down the finite difference scheme in the form of a linear system of algebraic equations. Solve the system of linear equations by the Gauss elimination method.*

Solution (a). If we assume that $u(x)$ is the unique theoretical solution of the boundary value problem (3.17) four times continuously differentiable, then $u(x)$ satisfies the difference equation

$$-\Lambda u_i + c_i u_i = f_i + \frac{h^2}{12}\frac{d^4u(\zeta_i)}{dx^4}, \qquad \zeta_i \in (x_{i-1}, x_{i+1}),$$

at mesh points $x_i = ih$, $i = 1, 2, ..., N-1$, $\quad Nh = l$.
Canceling the local truncation error

$$\psi_i(h) = \frac{h^2}{12}\frac{d^4u(\zeta_i)}{dx^4}$$

we arrive at the following $O(h^2)$ finite difference scheme:

$$-\Lambda v_i + c_i v_i = f_i, \qquad i = 1, 2, ..., N-1,$$

$$v_0 = \alpha, \qquad v_N = \beta. \tag{3.18}$$

Clearly, the local truncation error satisfies the inequality

$$|\psi_i(h)| \le \frac{h^2}{12}M^{(4)},$$

where the constant $M^{(4)} = \sup_{0 \le x \le l} |\frac{d^4u(x)}{dx^4}|$.

(b). Let $z_i = v_i - u_i$, $\ i = 0, 1, ..., N$, be the error of the method. Then, error $z \in H_h^0$ satisfies the following difference equation:

$$-\Lambda z_i + c_i z_i = \psi_i(h), \qquad i = 1, 2, ..., N-1,$$

$$z_0 = 0, \qquad z_N = 0. \tag{3.19}$$

Multiplying both sides of the equation (3.19) by $z \in H_h^0$, we obtain the following equation:

$$(-\Lambda z, z) + (cz, z) = (\psi, z).$$

By lemma 3.1 and by Cauchy-Buniakowsky inequality, we have

$$\frac{8}{l^2} \parallel z \parallel^2 + \mu \parallel z \parallel^2 \leq \parallel \psi \parallel \parallel z \parallel,$$

where $\mu = \inf_{0 \leq x \leq l} c(x) \geq 0$.

Thus, the error z holds inequality

$$\parallel z \parallel \leq \frac{M^{(4)} l^2}{12(\mu l^2 + 8)} h^2.$$

(c) Let us write the finite difference equation (3.18), when $c(x) = 0$, in the following form:

$$v_0 = \alpha$$

$$-v_{i-1} + 2v_i - v_{i+1} + h^2 c_i v_i = h^2 f_i, \quad i = 1, 2, ..., N - 1. \tag{3.20}$$

$$v_N = \beta.$$

The system of equations (3.20), we can write in the matrix form

$$\begin{bmatrix} 2 & -1 & 0 & \cdots & 0 & 0 \\ -1 & 2 & -1 & \cdots & 0 & 0 \\ \cdots & \cdots & \cdots & \cdots & \cdots & \cdots \\ 0 & 0 & 0 & \cdots & 2 & -1 \\ 0 & 0 & 0 & \cdots & -1 & 2 \end{bmatrix} \begin{bmatrix} v_1 \\ v_2 \\ \vdots \\ v_{N-2} \\ v_{N-1} \end{bmatrix} = h^2 \begin{bmatrix} f_1 + \alpha \\ f_2 \\ \vdots \\ f_{N-2} \\ f_{N-1} + \beta \end{bmatrix}. \tag{3.21}$$

Let us denote by

$$F_i = \begin{cases} h^2 f_i, & if \quad i = 1, N - 1, \\ h^2 f_i, & if \quad i = 2, 3, ..., N - 2 \end{cases}$$

Applying Gauss elimination to system of equations (3.21), one can obtain the following algorithm:

Algorithm:

$$Set :$$

$$\alpha_1 = -\frac{1}{2}, \qquad\qquad \beta_1 = \frac{F_1}{2}$$

$$evaluate : \quad for \ \ i = 2, 3, \ldots, N-1$$

$$\alpha_i = -\frac{1}{2 + \alpha_{i-1}}$$

$$\beta_i = \frac{F_i + \beta_{i-1}}{2 + \alpha_{i-1}} \tag{3.22}$$

$$set : \qquad\qquad x_{N-1} = \beta_{N-1}$$

$$evaluate : \quad for \ \ i = N-1, N-2, \ldots, 1$$

$$v_i = \frac{F_i + \beta_{i-1} + v_{i+1}}{2 + \alpha_{i-1}}.$$

Because

$$0 < |\alpha_i| < 1, \quad i = 1, 2, \ldots, N-1,$$

therefore, the coefficients $\alpha's$ move round-off errors from left to right decimal positions, so that round-off errors do not increase when the solution v_i, $i = 1, 2, \ldots, N-1$, is evaluated.

3.7   Discrete Maximum Principle.

The discrete maximum principle deals with finite difference operators of positive type. This class of operators is frequently associated with finite difference schemes for elliptic and parabolic equations. We shall consider the finite difference operators that are related with either one or two dimensional networks. Let

$$\overline{\Omega}_h = \{x_i = a + ih, \ \ i = 0, 1, \ldots, N, \ \ Nh = b - a\}$$

be one-parameter family of discrete sets in which a neighborhood $N(x_i)$ of a point x_i is defined. A neighborhood of a point can be introduced in different ways, for example, $N(x_i) = \{x_{i-1}, x_{i+1}\}$ is the two-point neighborhood of x_i or $N(x_i) = \{x_{i-2}, x_{i-1}, x_{i+1}, x_{i+2}\}$ is the four point neighborhood of x_i. Nevertheless, a neighborhood of x_i must be a non-empty subset of the set $\overline{\Omega}_h$. (Note that $x_i \notin N(x_i)$). We say, x_i is an interior point of $\overline{\Omega}_h$ if $N(x_i) \subset \overline{\Omega}_h$, and x_i is a boundary point of $\overline{\Omega}_h$ if $x_i \in \overline{\Omega}_h$, but $N(x_i)$ is not a subset of $\overline{\Omega}_h$. Consequently, the boundary $\partial\Omega_h$ of a network Ω_h is the set of all boundary points. Similarly, let

$$\overline{\Omega}_h = \{(x_i, y_j) \ : \ i = 0, 1, \ldots, N_1, \ j = 0, 1, \ldots, N_2, \ \ h = (h_1, h_2)\},$$

$$N_1 h_1 = b - a, \ \ N_2 h_2 = d - c.$$

be two-parameters family of sets in which the neighborhood $N(x_i, y_j)$ of a point (x_i, y_j) is defined. In two dimension, we can define the following neighborhood: $N(x_i, y_j) = \{x_{i\pm 1}, y_{j\pm 1})$ which is the four point neighborhood of (x_i, y_j). In a similar way, we say a point (x_i, y_j) is an interior point of $\overline{\Omega}_h$, if its neighborhood $N(x_i, y_j) \subset \overline{\Omega}_h$, and (x_i, y_j) is a boundary point of $\overline{\Omega}_h$ if $(x_i, y_j) \in \overline{\Omega}_h$, but $N(x_i, y_j)$ is not a subset of $\overline{\Omega}_h$. A discrete set Ω_h is called as a network if for each point of Ω_h a neighborhood is determined.

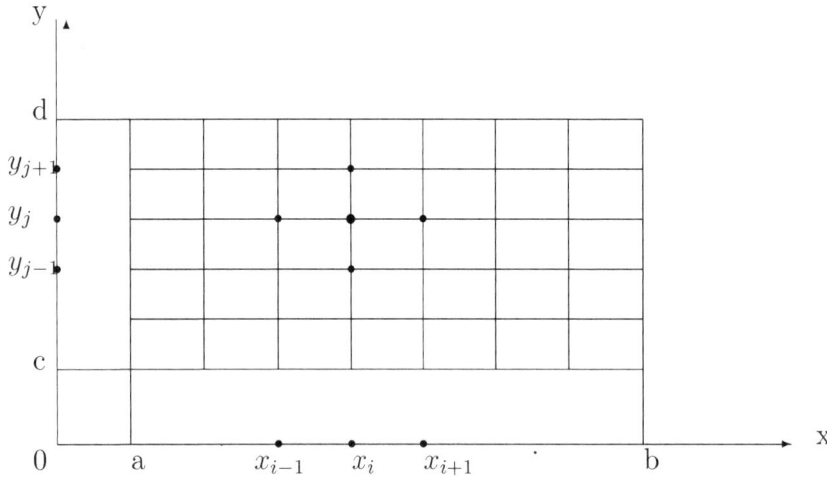

Fig. 3.2. Ω_h $-$ Rectangular network

Let us consider a linear operator

$$L_h[v](P) = a(P,P)v(P) + \sum_{Q \in N(P)} a(P,Q)v(Q), \qquad P \in \Omega_h, \qquad (3.23)$$

given at all discrete functions defined on network Ω_h, where the coefficients $a(P,Q)$ are also discrete functions for $P \in \Omega_h$ and $Q \in N(P)$.

3.8   Finite Difference Operators of Positive Type

Definition 3.2 *The linear operator L_h is said to be of positive type if and only if the following conditions hold:*

1. *(a) $a(P,Q) < 0$ for $P \neq Q$, $P \in \Omega_h$, $Q \in N(P)$,*
   where $N(P)$ is the neighbourhood of P,

   (b) $a(P,P) + \sum\limits_{Q \in N(P)} a(P,Q) \geq 0$ for all $P \in \Omega_h$,

   (c) there exists $P_0 \in \Omega_h$ such that $a(P_0, P_0) + \sum\limits_{Q \in N(P_0)} a(P_0, Q) > 0$

Example 3.3 *Consider the finite difference operator*

$$L_h v_i = -\Lambda v_i = -\frac{1}{h^2}[v_{i-1} - 2v_i + v_{i+1}],$$

defined at all discrete functions given on the network

$$\overline{\Omega}_h = \{0, h, 2h, ..., (N-1)h, Nh\}.$$

Obviously, L_h is a positive type operator. Indeed, L_h is a linear operator

$$L_h[v](P) = a(P, P)v(P) + \sum_{Q \in N(P)} a(P, Q)v(Q),$$

where $P = ih$ and the neighborhood $N(P) = \{(i-1)h, (i+1)h\}$. The coefficients

$$a(P, Q) = -\frac{1}{h^2} < 0 \quad \text{for} \quad P = ih \neq Q,$$

where $Q = (i-1)h$ or $Q = (i+1)h$.
Moreover, we have

$$a(P, P) + \sum_{Q \in N(P)} a(P, Q) = \begin{cases} -\dfrac{1}{h^2} + \dfrac{2}{h^2} - \dfrac{1}{h^2} = 0 & if \ P = 2h, 3h, ..., (N-2)h, \\[2mm] \dfrac{2}{h^2} - \dfrac{1}{h^2} = \dfrac{1}{h^2} > 0 & if \ P = h, (N-1)h. \end{cases}$$

3.8.1   The Maximum Princple

We shall consider the following discrete maximum principle:

The maximum principle 3.1 *(cf. [25],[29],[32]). Let L_h be a positive type operator. If v is a discrete function given on network $\overline{\Omega}_h$ which satisfies the inequality*

$$L_h[v](P) \leq 0 \quad \text{for} \quad P \in \Omega_h, \tag{3.24}$$

or

$$L_h[v](P) \geq 0 \quad \text{for} \quad P \in \Omega_h, \tag{3.25}$$

then

$$v(P) \leq \max\{0, \max_{Q \in \partial\Omega_h} v(Q)\}, \quad P \in \overline{\Omega}_h, \tag{3.26}$$

when (3.24) is true, or

$$v(P) \geq \min\{0, \min_{Q \in \partial\Omega_h} v(Q)\}, \quad P \in \overline{\Omega}_h, \tag{3.27}$$

when (3.25) is true.
Here $\partial\Omega_h$ is the boundary of the network $\Omega_h = \overline{\Omega}_h - \partial\Omega_h$.

Proof of the maximum principle. Let us assume that inequality (3.24) holds and, on the contrary, there exists an interior point $P_1 \in \Omega_h$ at which

$$\max_{Q \in \overline{\Omega}_h} v(Q) = v(P_1) \geq 0.$$

Then, by (a) and (b) the following inequalities hold:

$$
\begin{aligned}
L_h[v](P_1) &= a(P_1, P_1)v(P_1) + \sum_{Q \in N(P_1)} a(P_1, Q)v(Q) \\
&\geq v(P_1)[a(P_1, P_1) + \sum_{Q \in N(P_1)} a(P_1, Q)] \quad \geq \quad 0,
\end{aligned}
$$

or

$$
\begin{aligned}
L_h[v](P_1) &= a(P_1, P_1)v(P_1) + \sum_{Q \in N(P_1)} a(P_1, Q)v(Q) \\
&> v(P_1)[a(P_1, P_1) + \sum_{Q \in N(P_1)} a(P_1, Q)] \quad \geq \quad 0.
\end{aligned}
\tag{3.28}
$$

We observe that the strong inequality (3.28) holds if there exists at least one point $Q \in N(P_1)$ at which $v(Q) < v(P_1)$. However, in the case when $v(Q) < v(P_1)$, inequality (3.28) contradicts assumption (3.24) for certain $Q \in N(P_1)$. Therefore either thesis (3.26) is true or $v(Q) = v(P_1)$ for all $Q \in N(P_1)$. In the case when $v(Q) = v(P_1)$ for all $Q \in N(P_1)$, there exists a neighboring point $P_2 \in N(P_1)$ such that

$$P_2 \neq P_1 \quad \text{and} \quad v(P_2) = v(P_1).$$

If $P_2 \in \partial\Omega_h$ is a boundary point, then thesis (3.26) is true. Otherwise $P_2 \in \Omega_h$ is an interior mesh point and then, we can repeat the same analysis as carried out for the point P_1. Thus, one of two conclusions can be drawn, either thesis (3.26) is true or there exists an interior point $P_3 \in N(P_2)$ at which $v(P_3) = v(P_2) = v(P_1)$. We can continue the construction of points $P_1, P_2, P_3, ..., P_n$ until is met either a boundary point $P_n \in \partial\Omega_h$ or $v(P) = v(P_1)$ for all $P \in \Omega_h$. Thus, in both cases, inequality (3.26) is true. In a similar way, one can prove, inequality (3.27) under assumption (3.25).
The maximum principle implies the following:

Corolary 3.1 *Let v be the solution of the finite difference scheme*

$$
\begin{aligned}
L_h[v](P) &= 0, \qquad P \in \Omega_h, \\
v(P) &= \phi(P), \qquad P \in \partial\Omega_h,
\end{aligned}
\tag{3.29}
$$

where L_h is a positive type operator, $\phi(P)$ is a given discrete function on the boundary $\partial\Omega_h$ of the network Ω_h. Then, the following inequality holds:

$$\|v\|_{\overline{\Omega}_h} \leq \|\phi\|_{\partial\Omega_h},$$

where norms

$$||v||_{\overline{\Omega}_h} = \max_{P \in \overline{\Omega}_h} |v(P)|, \quad ||\phi||_{\partial \Omega_h} = \max_{P \in \partial \Omega_h} |\phi(P)|.$$

Indeed, then both assumptions (3.24) and (3.25) of the maximum principle hold. Therefore, by thesis (3.26), (3.27) solution v satisfies the inequality

$$\min\{0, \min_{Q \in \partial \Omega_h} v(Q)\} \leq v(P) \leq \max\{0, \max_{Q \in \partial \Omega_h} v(Q)\}, \quad P \in \overline{\Omega}_h.$$

Since

$$-||\phi||_{\partial \Omega_h} \leq \min\{0, \min_{Q \in \partial \Omega_h} v(Q)\}, \quad P \in \overline{\Omega}_h,$$

and

$$||\phi||_{\partial \Omega_h} \geq \max\{0, \max_{Q \in \partial \Omega_h} v(Q)\}, \quad P \in \overline{\Omega}_h,$$

we get

$$||v||_{\overline{\Omega}_h} \leq ||\phi||_{\partial \Omega_h}.$$

3.8.2   Stability in the Maximum Norm

In general, stability means continuous dependence of a solution on data. Thus, the finite difference scheme

$$L_h[v](P) = f(P), \quad P \in \Omega_h,$$

$$v(P) = \phi(P), \quad P \in \partial \Omega_h, \tag{3.30}$$

is stable if the solution v satisfies the following a priori estimate

$$||v||_{\overline{\Omega}_h} \leq K_1 ||\psi||_{\partial \Omega_h} + K_2 ||f||_{\Omega_h},$$

where K_1 and K_2 are constants independent of the step-size h, and $||.||_{\Omega_h}$, $||.||_{\partial \Omega_h}$ are norms which determine stability type. Let us note that stability in the norm $||.||$ of Hilbert's space H_h^0 means average convergence of a finite difference scheme and stability in the maximum norm $||.||_{\Omega_h}$ means uniform convergence of a finite difference scheme.

Let us prove the following theorem on stability in the maximum norm:

Theorem 3.1 *If L_h is a positive type operator and if there exists a discrete function $\beta_h(P)$ such that*

1.   *(a) $0 \leq \beta_h(P) \leq M$ for all $P \in \overline{\Omega}_h$,*
     (b) $L_h[\beta_h](P) \geq 1$ for all $P \in \Omega_h$,

then the finite difference scheme (3.30) is stable in the "maximum norm", i.e., solution v satisfies the inequality

$$||v||_{\overline{\Omega}_h} \leq ||\phi||_{\partial \Omega_h} + M ||f||_{\Omega_h},$$

where constant M is independent of h.

Proof. Let us introduce the following auxiliary discrete functions:

$$g(P) = ||\phi||_{\partial\Omega_h} + \beta_h(P)||f||_{\Omega_h},$$

$$w^-(P) = g(P) - v(P),$$

$$w^+(P) = g(P) + v(P).$$

By the assumptions (a) and (b)

$$
\begin{aligned}
L_h[w^\pm] &= L_h[g](P) - L_h[v](P) \\
&= [||\phi||_{\partial\Omega_h} L[1](P) + ||f||_{\Omega_h} L_h[\beta_h](P) \mp f(P) \\
&\geq ||\phi||_{\partial\Omega_h}[a(P,P) + \sum_{Q \in N(P)} a(P,Q)] + ||f||_{\Omega_h} \mp f(P) \\
&\geq ||f||_{\Omega_h} \mp f(P) \geq 0.
\end{aligned}
$$

The auxiliary functions $w^-(P)$ and $w^+(P)$ satisfy the following inequalities:

$$L_h[w^-](P) \geq 0, \qquad L_h[w^+](P) \geq 0 \quad \text{for} \quad P \in \Omega_h,$$

$$w^-(P) = g(P) - v(P) \geq 0, \qquad w^+(P) + v(P) \geq 0 \quad \text{for} \quad P \in \partial\Omega_h.$$

Hence, by the maximum principle

$$
\begin{aligned}
w^-(P) &\geq \min\{0, \min_{Q \in \partial\Omega_h} w^-(Q)\} \geq 0, \\
w^+(P) &\geq \min\{0, \min_{Q \in \partial\Omega_h} w^+(Q)\} \geq 0,
\end{aligned}
\tag{3.31}
$$

for all $P \in \overline{\Omega}_h$.

From inequalities (3.31) and definition of the auxiliary function $g(P)$

$$g(P) \pm v(P) \geq 0 \quad \text{for} \quad P \in \overline{\Omega}_h,$$

or

$$-g(P) \leq v(P) \leq g(P) \quad \text{for} \quad P \in \overline{\Omega}_h.$$

Hence

$$g(P) \leq ||\phi||_{\partial\Omega_h} + M||f||_{\Omega_h}, \quad P \in \overline{\Omega}_h,$$

and by the definition of $g(P)$, we obtain the following a priori estimate:

$$||v||_{\overline{\Omega}_h} \leq ||\phi||_{\partial\Omega_h} + M||f||_{\Omega_h}.$$

which means stability of (3.30) in the *maximum norm* .

Example 3.4 *Show that the finite difference scheme*

$$-\Lambda v_i + c_i v_i = f_i, \quad i = 1, 2, ..., N - 1,$$

$$v_0 = B, \qquad v_N = C,$$

$$\tag{3.32}$$

is stable in the maximum norm, provided that $c_i \geq 0$, for $i = 1, 2, ..., N - 1$.

Solution. In order to show stability of the finite difference scheme (3.32), we shall give the following a priori estimate of the solution v:

$$||v||_{\overline{\Omega}_h} \leq \max\{|B|, |C|\} + 2||f||_{\Omega_h}.$$

Let us note that the discrete function

$$\beta_h(ih) = 3 - (1+h)^i, \quad i = 0, 1, ..., N,$$

satisfies the assumptions of the theorem. Indeed, we have

$$0 \leq 3 - (1+h)^i \leq 2, \quad i = 0, 1, ..., N,$$

and

$$
\begin{aligned}
L_h[\beta_h](ih) &= -\Lambda[\beta_h](ih) + c_i\beta_h(ih) \\
&= -\frac{1}{h^2}[\beta_h((i-1)h) - 2\beta_h(ih) + \beta_h((i+1)h)] + c_i\beta_h(ih) \\
&\geq -\frac{1}{h^2}[3 - (1+h)^{i-1} - 2(3 - (1+h)^i + 3 - (1+h)^{i+1}] \\
&= (1+h)^{i-1} \geq 1,
\end{aligned}
$$

for $i = 1, 2, ..., N-1$.
By theorem (3.1)

$$||v||_{\overline{\Omega}_h} \leq \max\{|B|, |C|\} + 2||f||_{\Omega_h}.$$

3.9   Exercises

Question 3.1 *Solve the following eigenvalue problem:*

1. *(a)*
$$-2\Lambda v_i + 3v_i = \lambda v_i, \quad i = 1, 2, ..., N-1, \ Nh = 4,$$

$$v_0 = 0, \quad v_n = 0.$$

   (b) Find the lower-bound of the eigenvalues λ_k, $k = 1, 2, ..., N-1$.
   (c) Show that the operator

$$L_h v_i = -\Lambda v_i + 3v_i, \quad i = 1, 2, ..., N-1,$$

   is positive definite in the Hilbert space H_h^0.

Question 3.2 *Show that the eigenfunctions $v^{(k)}$, $k = 1, 2, ..., N-1$, $Nh = 4$, of the finite difference operator*

$$L_h = -2\Lambda + 3I$$

are orthogonal in the Hilbert space H_h^0, where I is the identity operator. Find the norm $||v^{(k)}||$, $k = 1, 2, ..., N-1$.

Question 3.3 *Consider the following finite difference scheme*

$$-3\Lambda v_i + 4v_i = \frac{ih}{1+ih}, \quad i = 1, 2, ..., N-1, \quad Nh = 2,$$

$$v_0 = -2, \quad v_N = 2.$$

Estimate the solution v in the norm of the Hilbert space H_h^0.

Question 3.4 *Consider the following boundary value problem:*

$$-3\frac{d^2 u(x)}{dx^2} + 5u(x) = \exp(-x), \quad 0 \le x \le 2, \tag{3.33}$$

$$u(0) = 1, \quad u(2) = 0.$$

1. (a) *Approximate the boundary value problem (3.33) by a finite difference scheme of the second order. Determine the truncation error.*

    (b) *Estimate the error of the method in the norm of the Hilbert space H_h^0.*

Question 3.5 *Consider the following finite difference scheme:*

$$-2\Lambda v_i + 3v_i = \frac{ih}{1+ih}, \quad i = 1, 2, ..., N-1,$$

$$v_0 = -3, \quad v_N = 5, \quad Nh = 2,$$

Give an a priori estimate of solution v in the maximum norm.

Question 3.6 *Show that the finite difference scheme*

$$-3\Lambda v_i + 2v_i = \sin \pi\, ih, \quad i - 1, 2, ..., N-1,$$

$$v_0 = 1, \quad v_N = 4, \quad Nh = 2,$$

is stable in the maximum norm.

Question 3.7 *Approximate the boundary value problem*

$$-3\frac{d^2 u(x)}{dx^2} + 2u(x) = \sin \pi\, x, \quad 0 \le x \le 2,$$

$$u(0) = 2, \quad u(2) = 1,$$

by an $O(h^2)$ convergent finite difference scheme. Estimate the error of the method in the maximum norm.

Question 3.8 *Consider the following boundary value problem:*

$$-\frac{d^2 u(x)}{dx^2} = e^{-u(x)}, \quad 0 \le x \le 1,$$

$$u(0) = 0, \quad u(1) = 0.$$

1. (a) Use `Mathematica` *module* `solveBVP` *to solve the boundary value problem.*

   (b) *Give the approximate solution in the form of an interpolating polynomial.*

   (c) *Tabulate the residual error using step size* $h = 0.05$.

Chapter 4

Elliptic Equations

Tadeusz Styš
University of Warsaw

Abstract: In chapter 4, two finite difference schemes are built for Helmholtz equation with Dirichlet's boundary value conditions. It is proved that the first scheme is $O(h_1^2 + h_2^2)$ convergent in the norm of the Hilbert's space H and in the maximum norm. The second scheme is $O(h_1^4 + h_2^4)$ accurate and is also convergent in the norm of the Hilbert's space H and in the maximum norm. Both schemes are solved by the *Mathematica* module. In the last section, Poisson's equation with Dirichlet's boundary conditions is solved by the method of lines. The chapter ends with a set of questions.

Chapter 4

Elliptic Equations

4.1   Introduction

In general, elliptic equations are associated with mathematical models of equilibrium problems. We shall deal with linear and nonlinear elliptic equations. Linear elliptic equations are most frequently represented by Poisson and Helmholtz equations. We shall use these two equations in describing the finite difference method in its standard form. More technically developed analysis of convergence concerns n-dimensional nonlinear elliptic equations and elliptic systems of equations. For such equations, we shall also use representative models to simplify analysis and notation.

In order to approximate an elliptic equation by the finite difference method, we first define mesh points and then classify them in two groups:

- interior mesh points at which the difference equation is satisfied,

- boundary mesh point at which boundary conditions hold.

Let us define the rectangular network Ω_h with the interior mesh points

$$\Omega_h = \{(ih_1, jh_2) \; : \; i = 1, 2, ..., N_1 - 1; \quad j = 1, 2, ..., N_2 - 1\}$$

and with the boundary mesh points

$$\partial\Omega_h = \{(ih_1, jh_2) \; : \; i = 0, N_1, \; j = 0, 1, ..., N_2, \; or \; j = 0, N_2, \; i = 0, 1, ..., N_1\},$$

where h_1 and h_2 are determined by the equalities $N_1 h_1 = l_1, \quad N_2 h_2 = l_2$.

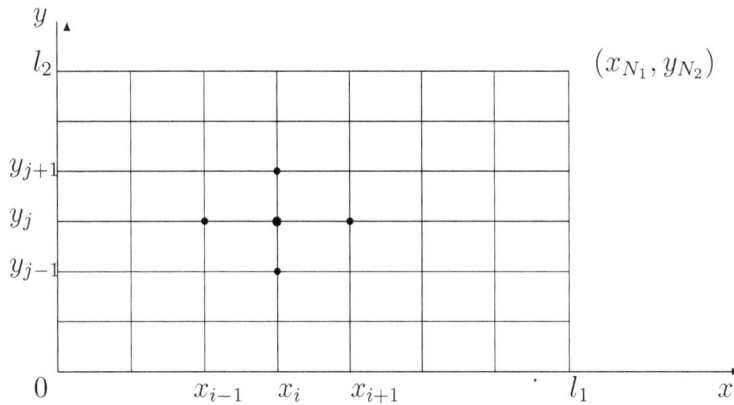

Fig. 4.1. Rectangular network Ω_h.

Let $u(x, y)$ be a function four times continuously differentiable in the rectangle $R = \{(x, y) : 0 \le x \le l_1, 0 \le y \le l_2\}$ and let $u_{ij} = u(x_i, y_j)$ be the value of u at the point (x_i, y_j), where $x_i = ih_1, \quad y_j = jh_2$. It is then easy to show that the second partial derivatives of u satisfy the following relations:

$$\Lambda_x u_{ij} \equiv \frac{u_{i-1j} - 2u_{ij} + u_{i+1j}}{h_1^2} = \frac{\partial^2 u_{ij}}{\partial x^2} + \frac{h_1^2}{12} \frac{\partial^4 u(\xi_i, y_j)}{\partial x^4},$$
$$\xi \in (x_{i-1}, x_{i+1}), \qquad (4.1)$$
$$\Lambda_y u_{ij} \equiv \frac{u_{ij-1} - 2u_{ij} + u_{ij+1}}{h_2^2} = \frac{\partial^2 u_{ij}}{\partial y^2} + \frac{h_2^2}{12} \frac{\partial^4 u(x_i, \zeta_j)}{\partial y^4},$$

for all interior mesh points $(ih_1, jh_2) \in \Omega_h$.

Thus, the partial derivatives $\dfrac{\partial^2 u_{ij}}{\partial x^2}, \dfrac{\partial^2 u_{ij}}{\partial y^2},$ which appear in an equation can be replaced by $\Lambda_x u_{ij}, \Lambda_y u_{ij}$ with the local truncation errors

$$\psi_{ij}^1(h_1) \frac{h_1^2}{12} \frac{\partial^4 u(\xi_i, y_j)}{\partial x^4}, \quad \psi_{ij}^2(h_2) = \frac{h_2^2}{12} \frac{\partial^4 u(x_i, \zeta_j)}{\partial y^4}.$$

for certain $\xi_i \in (x_{i-1}, x_{i+1})$ and $\zeta_j \in (y_{j-1}, y_{j+1})$.

4.2   Linear Elliptic Equations

Let us consider the Helmholtz equation

$$-\frac{\partial^2 u(x, y)}{\partial x^2} - \frac{\partial^2 u(x, y)}{\partial y^2} + c(x, y)u(x, y) = f(x, y), \quad (x, y) \in R, \qquad (4.2)$$

with the Dirichlet boundary value condition

$$u(x, y) = \phi(x, y), \quad (x, y) \in \partial R. \tag{4.3}$$

We assume that $c(x, y) \geq 0$, $f(x, y)$ and $\phi(x, y)$ are given continuous functions on the rectangle R and its boundary ∂R, respectively.

Then, the theoretical solution $u(x, y)$ satisfies the difference equation

$$-\Lambda_x u_{ij} - \Lambda_y u_{ij} + c_{ij} u_{ij} = f_{ij} + \psi_{ij}(h_1, h_2), \quad (ih_1, jh_2) \in \Omega_h, \tag{4.4}$$

where the local truncation error is

$$\psi_{ij}(h_1, h_2) = \frac{h_1^2}{12} \frac{\partial^4 u(\xi_i, y_j)}{\partial x^4} + \frac{h_2^2}{12} \frac{\partial^4 u(x_i, \zeta_j)}{\partial y^4}, \tag{4.5}$$

This error satisfies the following inequality:

$$|\psi_{ij}(h_1, h_2)| \leq \frac{h_1^2 + h_2^2}{12} M^{(4)}, \tag{4.6}$$

where $M^{(4)} = \max\{\sup |\frac{\partial^4 u(x, y)}{\partial x^4}|, \sup |\frac{\partial^4 u(x, y)}{\partial y^4}|\}$.

By canceling the truncation error $\psi_{ij}(h_1, h_2)$ in equation (4.4), we obtain the following $O(h_1^2 + h_2^2)$ accurate five point finite difference scheme:

$$-\Lambda_x v_{ij} - \Lambda_y v_{ij} + c_{ij} v_{ij} = f_{ij}, \quad (ih_1, jh_2) \in \Omega_h,$$
$$v_{ij} = \phi_{ij}, \quad\quad\quad\quad (ih_1, jh_2) \in \partial\Omega_h. \tag{4.7}$$

The finite difference scheme (4.7) constitutes a system of linear equations that can be solved by an iterative method. Below, we write this system of linear equations in the form of the following module:

Program 4.1 *Mathematica module for discrete Helmholtz equation*

```
discreteHelmholtzEqn[n1_,n2_]:=Module[{cv, m, row0, lsh, lsh1,
        lsh2, u, umatrix },
        Clear[u,h1,h2]; row0=Table[0,{n2+1}];
        umatrix=Prepend[Append[
                Map[Prepend[Append[#,0],0]&,
                Array[u,{n1-1,n2-1}]], row0], row0];
        TableForm[umatrix];
        laplace1[v_, h_]:=
                (RotateRight[v,1] -2 v +RotateLeft[v,1])/h^2;
        lhs1=Transpose[Map[laplace1[#, h1]&,
```

```
                         Transpose[umatrix]]];
             lhs2=Map[laplace1[#,h2]&,umatrix];
             lhs=Take[Map[Take[#, {2,-2}]&, lhs1+lhs2], {2,-2}];
             cv=MapThread[Times,{Array[c,{n1-1,n2-1}],
                          Array[u,{n1-1,n2-1}]}];
             m=Array[f,{n1-1,n2-1}];
             MapThread[Equal,{-lhs+cv,m},2]
]
```

The module `discreteHelmholtzEqn` creates in `Mathematica` the finite difference scheme (4.7) with the homogeneneous boundary value conditions for the
number of interior mesh points $(n1 - 1) * (n2 - 1)$ and given functions c_{ij}, f_{ij}.

Example 4.1 *Let us consider the Poisson's equation*

$$-\frac{\partial^2 u}{\partial x^2} - \frac{\partial^2 u}{\partial y^2} = 20(x(1 - x) + y(1 - y)), \quad 0 \le x, y \le 1,$$

with the homogeneous boundary conditions.

Solving the boundary value problem with `Mathematica` by Jacobi's iterative
method, we can use the following `module`:

Program 4.2 *Mathematica module that solves discrete Helmholtz equation*

```
jacobHelmholtzEqn[u_,a_,b_,c_,f_]:=Module[
            {d,m,n1,n2,h1,h2,lhs,lhs1,lhs2,cv,row0,sol,laplace1},

            {n1,n2}=Dimensions[u]-1;
            h1=(b-a)/n1; h2=(b-a)/n2;
            row0=Table[0,{n2+1}];

            laplace1[v_, h_]:=
            (RotateRight[v,1]+RotateLeft[v,1])/h^2;
            lhs1=Transpose[Map[laplace1[#, h1]&,Transpose[u]]];
            lhs2=Map[laplace1[#,h2]&,u];
            lhs=Take[Map[Take[#, {2,-2}]&, lhs1+lhs2], {2,-2}];
            cv=Table[c[i*h1,j*h2],{i,1,n1-1},{j,1,n2-1}];
            m=Table[f[i*h1,j*h2],{i,1,n1-1},{j,1,n2-1}];
            d=2/h1^2+2/h2^2;
            sol=(m+lhs)/(d+cv);
            Prepend[Append[Map[Prepend[Append[#,0],0]&,sol],
             row0], row0]
]
```

In order to invole the module `jacobHelmholtzEqn`, we enter input data

```
m1=4;       m2=6;       h1=1/m1;       h2=1/m2;
f[x_,y_]:=20*(x*(1-x)+y*(1-y));
c[x_,y_]:=0;
uzero=Table[0,{m1+1},{m2+1}];
```

Then, we execute the `Nest` command

```
N[Nest[jacobHelmholtzEqn[#,0,1,c,f]&,uzero,40],3]//TableForm
```

to obtain the following approximate solution

0	0	0	0	0	0	0
0	0.26	0.417	0.469	0.417	0.26	0
0	0.347	0.555	0.625	0.555	0.347	0
0	0.26	0.417	0.469	0.417	0.26	0
0	0	0	0	0	0	0

Let us note that the theoretical solution is $u(x, y) = 10x(1 - x)y(1 - y)$.
$O(h_1^4 + h_2^4)$ **Finite Difference Scheme.** We shall now deal with Poisson's
equation

$$-\frac{\partial^2 u(x, y)}{\partial x^2} - \frac{\partial^2 u(x, y)}{\partial y^2} = f(x, y), \quad (x, y) \in R, \tag{4.8}$$

and with Dirichlet's boundary value condition

$$u(x, y) = \phi(x, y), \quad (x, y) \in \partial R.$$

If $u(x, y)$ is six times continuously differentiable of (4.8), then

$$
\begin{aligned}
-\Lambda_x u_{ij} - \Lambda_y u_{ij} = {} & f_{ij} - \frac{h_1^2}{12}\frac{\partial^4 u_{ij}}{\partial x^4} - \frac{h_2^2}{12}\frac{\partial^4 u_{ij}}{\partial y^4} \\
& - \frac{h_1^4}{360}\frac{\partial^6 u(\xi_i, jh_2)}{\partial x^6} - \frac{h_2^4}{360}\frac{\partial^6 u(ih_1, \zeta_j)}{\partial y^6},
\end{aligned}
\tag{4.9}
$$

for certain $\xi_i \in ((x_{i-1}h_1, x_{i+1}h_1)$ and $\zeta_j \in (y_{j-1}h_2, y_{j+1}h_2)$.
Since

$$\frac{\partial^4 u_{ij}}{\partial x^4} = -\frac{\partial^2 f_{ij}}{\partial x^2} - \frac{\partial^4 u_{ij}}{\partial x^2 \partial y^2} \quad \text{and} \quad \frac{\partial^4 u_{ij}}{\partial y^4} = -\frac{\partial^2 f_{ij}}{\partial y^2} - \frac{\partial^4 u_{ij}}{\partial x^2 \partial y^2},$$

we have

$$
\begin{aligned}
-\Lambda_x u_{ij} - \Lambda_y u_{ij} = \ & f_{ij} + \frac{h_1^2}{12}\frac{\partial^2 f_{ij}}{\partial x^2} + \frac{h_2^2}{12}\frac{\partial^2 f_{ij}}{\partial y^2} + \frac{h_1^2 + h_2^2}{12}\frac{\partial^4 u_{ij}}{\partial x^2 \partial y^2} \\
& - \frac{h_1^4}{360}\frac{\partial^6 u(\xi_i, jh_2)}{\partial x^6} - \frac{h_2^4}{360}\frac{\partial^6 u(ih_1, \zeta_j)}{\partial y^6}.
\end{aligned}
\tag{4.10}
$$

Let us substitute the following into (4.10)

$$
\begin{aligned}
\frac{\partial^2 f_{ij}}{\partial x^2} &= \Lambda_x f_{ij} - \frac{h_1^2}{12}\frac{\partial^4 f(\xi_i, jh_2)}{\partial x^4}, \\
\frac{\partial^2 f_{ij}}{\partial y^2} &= \Lambda_y f_{ij} - \frac{h_2^2}{12}\frac{\partial^4 f(ih_1, \zeta_j)}{\partial y^4}, \\
\frac{\partial^4 u_{ij}}{\partial x^2 \partial y^2} &= \Lambda_x \Lambda_y u_{ij} - \frac{h_1^2}{12}\frac{\partial^6 u(\xi_i, jh_2)}{\partial x^4 \partial y^2} - \frac{h_2^2}{12}\Lambda_x\Big[\frac{\partial^4 u(ih_1, \zeta_j)}{\partial y^4}\Big].
\end{aligned}
$$

Then, we obtain

$$
\begin{aligned}
-\Lambda_x u_{ij} - \Lambda_y u_{ij} \ = \ & f_{ij} + \frac{h_1^2}{12}\Lambda_x f_{ij} + \frac{h_2^2}{12}\Lambda_y f_{ij} \\
& + \frac{h_1^2 + h_2^2}{12}\Lambda_x \Lambda_y u_{ij} + \psi_{ij}(h_1, h_2),
\end{aligned}
\tag{4.11}
$$

where the local truncation error is

$$
\begin{aligned}
\psi_{ij}(h_1, h_2) = \ & -\frac{h_1^4}{144}\frac{\partial^4 f(\xi_i, jh_2)}{\partial x^4} - \frac{h_2^4}{144}\frac{\partial^4 f(ih_1, \zeta_j)}{\partial x^4} \\
& - \frac{(h_1^2 + h_2^2)h_1^2}{144}\frac{\partial^6 u(\xi_i, jh_2)}{\partial x^4 \partial y^2} - \frac{(h_1^2 + h_2^2)h_1^2}{144}\Lambda_x\Big[\frac{\partial^4 u(ih_1, \zeta_j)}{\partial y^4}\Big] \\
& - \frac{h_1^4}{360}\frac{\partial^6 u(\xi_i, jh_2)}{\partial x^6} - \frac{h_2^4}{360}\frac{\partial^6 u(ih_1, \zeta_j)}{\partial y^6} = O(h_1^4 + h_2^4).
\end{aligned}
$$

By canceling the truncation error in (4.11), we get $O(h_1^4 + h_2^4)$ finite difference scheme

$$
\begin{aligned}
-\Lambda_x v_{ij} - \Lambda_y v_{ij} - \frac{h_1^2 + h_2^2}{12}\Lambda_x \Lambda_y v_{ij} \ &= \\
= \frac{f_{i-1j} + f_{i+1j} + f_{ij-1} + f_{ij+1} - 8f_{ij}}{12}, \quad & (ih_1, jh_2) \in \Omega_h, \\
v_{ij} = \phi_{ij}, \quad & (ih_1, jh_2) \in \partial\Omega_h.
\end{aligned}
\tag{4.12}
$$

In fact, the construction of finite difference schemes for boundary value problems is a straight forward procedure. However, analysis of convergence and an estimate of the global error require more developed techniques. In the next sections, we shall introduce two techniques frequently used in analysis of convergence of the finite difference method for elliptic equations. The first technique is based on the Hilbert space properties and it leads to average convergence in the discrete norm of the $L_2(R)$ space. The second one concerns the discrete maximum principle and it is used to prove uniform convergence of the method.

4.2.1 Average Convergence in the Hilbert Space $H_h^0(\Omega_h)$

Let us consider the set $H_h^0(\Omega_h)$ of all discrete functions given on the network Ω_h which vanish at the boundary $\partial\Omega_h$. In a way similar to what is done for one variable functions, we introduce the following inner product in $H_h^0(\Omega_h)$:

$$(v, w) = h_1 h_2 \sum_{i=1}^{N_1-1} \sum_{j=1}^{N_2-1} v_{ij} w_{ij}, \quad v, w \in H_h^0(\Omega_h). \qquad (4.13)$$

The set $H_h^0(\Omega_h)$ with the inner product $(-,-)$ is the Hilbert space $H_h^0(\Omega_h)$. The inner product (4.13) generates the norm

$$||v|| = \sqrt{(v, v)} \quad \text{for each} \quad v \in H_h^0(\Omega_h).$$

Eigenvalue problem. *Find all numbers λ and all non-zero discrete functions $v \in H_h^0(\Omega_h)$, corresponding to λ such that*

$$-\Lambda v_{ij} \equiv -\Lambda_x v_{ij} - \Lambda_y v_{ij} = \lambda v_{ij}, \quad (ih_1, jh_2) \in \Omega_h,$$
$$v_{ij} = 0, \qquad (ih_1, jh_2) \in \partial\Omega_h. \qquad (4.14)$$

Solution. In order to solve the eigenvalue problem, we apply the method of separation of variables. Then, we predict solution v in the form of the product

$$v_{ij} = W_i T_j, \quad i = 0, 1, ..., N_1, \ j = 0, 1, ..., N_2$$

of two discrete functions W_i and T_j. Each of these functions depends either on i or j.
By substituting $v_{ij} = W_i T_j$ into (4.14), we get

$$-\Lambda v_{ij} = T_j(-\Lambda_x W_i) + W_i(-\Lambda_y T_j) = \lambda W_i T_j,$$

for $i = 1, 2, ..., N_1 - 1, \ j = 1, 2, ..., N_2 - 1$.
Hence

$$-\frac{\Lambda_x W_i}{W_i} - \frac{\Lambda_y T_j}{T_j} = \lambda, \quad (ih_1, jh_2) \in \Omega_h,$$

or

$$-\frac{\Lambda_x W_i}{W_i} = \frac{\Lambda_y T_j}{T_j} + \lambda, \quad (ih_1, jh_2) \in \Omega_h.$$

Therefore

$$-\frac{\Lambda_x W_i}{W_i} = \text{constant} = \mu \;\; \text{and} \;\; \frac{\Lambda_y T_j}{T_j} + \lambda = \mu, \quad (ih_1, jh_2) \in \Omega_h.$$

In this way, we have split ted the eigenvalue problem (4.14) into the following one-dimensional problems:

$$-\Lambda_x W_i = \mu W_i, \quad i = 1, 2, ..., N_1 - 1,$$

$$W_0 = 0, \quad W_{N_1} = 0,$$

(4.15)

and

$$-\Lambda_y T_j = \nu T_j, \quad j = 1, 2, ..., N_2 - 1,$$

$$T_0 = 0, \quad T_{N_2} = 0.$$

(4.16)

where $\lambda = \mu + \nu$.

As we know, the solution of eigenvalue problem (4.15) is:

$$\mu_k = \frac{4}{h_1^2} \sin^2 \frac{k\pi h_1}{2l_1}, \quad k = 1, 2, ..., N_1 - 1,$$

$$W_i^{(k)} = \sqrt{\frac{2}{l_1}} \sin \frac{k\pi i h_1}{l_1}, \quad i = 0, 1, ..., N - 1,$$

(4.17)

and, similarly, the solution of eigenvalue problem (4.16) is:

$$\nu_m = \frac{4}{h_2^2} \sin^2 \frac{m\pi h_2}{2l_2}, \quad m = 1, 2, ..., N_1 - 1,$$

$$T_j^{(m)} = \sqrt{\frac{2}{l_2}} \sin \frac{m\pi j h_2}{l_2}, \quad j = 0, 1, ..., N_2 - 1.$$

(4.18)

Combining these solutions, we find the eigenvalues of $-\Lambda$

$$\lambda_{km} = \mu_k + \nu_m = \frac{4}{h_1^2} \sin^2 \frac{k\pi h_1}{2l_1} + \frac{4}{h_2^2} \sin^2 \frac{m\pi h_2}{2l_2},$$

and the eigenfunctions of $-\Lambda$

$$v_{ij}^{(km)} = \sqrt{\frac{4}{l_1 l_2}} \sin \frac{k\pi i h_1}{l_1} \sin \frac{m\pi j h_2}{l_2},$$

where $k = 1, 2, ..., N_1 - 1$, $m = 1, 2, ..., N_2 - 1$, $i = 0, 1, ..., N_1$, $j = 0, 1, ..., N_2$. The set $\{v^{(km)}\}$, $k = 1, 2, ..., N_1 - 1$, $m = 1, 2, ..., N_2 - 1$, of eigenfunctions is orthonormal in the Hilbert space $H_h^0(\Omega_h)$, i.e.,

$$(v^{(km)}, v^{(lp)}) = \begin{cases} 1 & if \;\; k = l \;\; and \;\; m = p, \\ \\ 0 & otherwise. \end{cases}$$

(4.19)

Indeed, we have

$$
\begin{aligned}
\left(v^{(km)}, v^{lp}\right) &= h_1 h_2 \sum_{i=1}^{N_1-1} \sum_{j=1}^{N_2-1} v_{ij}^{km} v_{ij}^{lp} \\
&= \{h_1 \sum_{i=1}^{N_1-1} W_i^{(k)} W_i^{(l)}\}\{h_2 \sum_{j=1}^{N_2-1} T_j^{(m)} T_j^{(p)}\} \\
&= (W^{(k)}, W^{(l)})(T^{(m)}, T^{(p)}) = \delta_{kl}\delta_{mp},
\end{aligned}
$$

where δ_{kl} and δ_{mp} are Kronecker's δ.

As in one dimension, it can be shown that the operator Λ is symmetric in the Hilbert space $H_h^0(\Omega_h$, that is

$$(\Lambda v, w) = (v, \Lambda w) \text{ for any } v, w \in H_h^0(\Omega_h).$$

The following lemma holds:

Lemma 4.1 *Every discrete function* $v \in H_h^0(\Omega_h)$ *satisfies the inequality:*

$$(\frac{8}{l_1^2} + \frac{8}{i_2^2})||v||^2 \le (-\Lambda v, v) \le (\frac{4}{h_1^2} + \frac{4}{h_2^2})||v||^2. \tag{4.20}$$

Proof of lemma (4.1). Let $v \in H_h^0(\Omega_h)$. Then, by lemma (3.1) the discrete function $v_{*j} = (0, v_{1j}, v_{2j}, ..., v_{N_1-1j}, 0)$ satisfies the inequality

$$\frac{8}{h_1^2}||v_{*j}||^2 \le (-\Lambda_x v_{*j}, v_{*j}) \le \frac{4}{h_1^2}||v_{*j}||^2, \tag{4.21}$$

for $j = 1, 2, ..., N_2 - 1$.

Also, by lemma (3.1), the discrete function $v_{i*} = (0, v_{i1}, v_{i2}, ..., v_{iN_2-1}, 0)$ satisfies the inequality

$$\frac{8}{h_2^2}||v_{i*}||^2 \le (-\Lambda_x v_{i*}, v_{i*}) \le \frac{4}{h_1^2}||v_{i*}||^2. \tag{4.22}$$

for $i = 1, 2, ..., N_1 - 1$,

Let us note that

$$||v||^2 = (v, v) = h_1 h_2 \sum_{i=1}^{N_1-1} \sum_{j=1}^{N_2-1} v_{ij}^2 = \sum_{j=1}^{N_2-1} h_2(v_{*j}, v_{*j}) = h_2 \sum_{j=1}^{N_2-1} ||v_{*j}||^2,$$

$$||v||^2 = (v, v) = h_1 h_2 \sum_{i=1}^{N_1-1} \sum_{j=1}^{N_2-1} v_{ij}^2 = \sum_{i=1}^{N_1-1} h_1(v_{i*}, v_{i*}) = h_1 \sum_{i=1}^{N_1-1} ||v_{i*}||^2,$$

and

$$(-\Lambda v, v) = h_1 h_2 \sum_{i=1}^{N_1-1} \sum_{j=1}^{N_2-1} [v_{ij}(-\Lambda_x v_{ij}) + v_{ij}(-\Lambda_y v_{ij})] = (-\Lambda_x v, v) + (-\Lambda_y v, v).$$

Hence, by inequalities (4.21) and (4.22), we obtain the expected inequality (4.20).

From lemma (4.1) it follows that operator $-\Lambda$ is positive definite in the Hilbert space $H_h^0(\Omega_h)$. Indeed, $-\Lambda$ is a symmetric operator, and by lemma (4.1)

$$(-\Lambda v, v) \geq \gamma(v, v), \quad \text{for all} \quad v \in H_h^0(\Omega_h),$$

where $\gamma = \dfrac{8}{l_1^2} + \dfrac{8}{l_2^2}$.

Thus, by definition (3.1), $-\Lambda$ is a positive definite operator .

Convergence in $H_h^0(\Omega_h)$. Let us now prove convergence of finite difference schemes (4.7) and (4.12).

One can check that if v is the solution of (4.7), then the global error of the method $z = v - u$ satisfies the difference equation

$$-\Lambda z_{ij} + c_{ij} z_{ij} = \psi_{ij}(h_1, h_2), \quad (ih_1, jh_2) \in \partial\Omega_h,$$

$$z_{ij} = 0, \quad (ih_1, jh_2) \in \partial\Omega_h. \tag{4.23}$$

where $\psi_{ij}(h_1, h_2)$, $i = 1, 2, ..., N_1 - 1$, $j = 1, 2, ..., N_1 - 1$, is the local truncation error given by (4.5).

Obviously, the error z vanishes at the boundary $\partial\Omega_h$ of the network Ω_h and therefore $z \in H_h^0(\Omega_h)$. Multiplying both sides of the scheme (4.23) by z in the sense of the inner product $(-, -)$, we obtain the following equation:

$$(-\Lambda z, z) + (cz, z) = (\psi, z). \tag{4.24}$$

By lemma (4.1) and by Cauchy's inequality

$$\gamma \, ||z||^2 + c_{min} ||z||^2 \leq ||\psi|| \, ||z||,$$

where $\gamma = \dfrac{8}{l_1^2} + \dfrac{8}{l_2^2}$ and $c_{min} = \min \, c(x, y) \geq 0$.

Hence, by (4.6), z satisfies the following inequality:

$$||z|| \leq \frac{||\psi||}{\gamma + c_{min}} \leq \frac{h_1^2 + h_2^2}{12(\gamma + c_{min})} \, M^{(4)}.$$

Thus, the error $||z|| = ||v - u|| \to 0$ as fast as $h_1^2 + h_2^2 \to 0$, and the mean value of $z = O(h_1^2 + h_2^2)$.

Similarly, if v is the solution of (4.12), then the global error $z = v - u \in H_h^0(\Omega_h)$ and

$$-\Lambda_x z_{ij} - \Lambda_y z_{ij} - \frac{h_1^2 + h_2^2}{12} \Lambda_x \Lambda_y z_{ij} = \psi_{ij}(h_1, h_2), \tag{4.25}$$

for $(ih_1, jh_2) \in \Omega_h$, where $\psi_{ij}(h_1, h_2) = O(h_1^4 + h_2^4)$.

Multiplying (4.25) by z, we have

$$(-L_h z, z) = (\psi, z), \tag{4.26}$$

where the operator

$$L_h = \Lambda_x + \Lambda_y + \frac{h_1^2 + h_2^2}{12}\Lambda_x\Lambda_y.$$

Because

$$(-\Lambda_x z, z) \geq \frac{8}{l_1^2}||z||^2 \quad \text{and} \quad (-\Lambda_y z, z) \geq \frac{8}{l_2^2}||z||^2,$$

we have

$$(\Lambda_x\Lambda_y z, z) \geq \frac{64}{l_1^2 l_2^2}$$

and

$$(-L_h z, z) \geq [\frac{8}{l_1^2} + \frac{8}{l_2^2} - \frac{64}{l_1^2 l_2^2}]||z||^2.$$

Hence, for $h_1 \leq \dfrac{l_1}{4}$ and $h_2 \leq \dfrac{l_2}{4}$, we get

$$(-L_h z, z) \geq \gamma||z||^2, \tag{4.27}$$

where $\gamma = \dfrac{4}{l_1^2} + \dfrac{4}{l_2^2}$.

From (4.26) and (4.27), and by Cauchy's inequality

$$||z|| \leq \frac{1}{\gamma}||\psi||.$$

Thus, the norm of the local truncation error

$$||\psi|| \leq K(h_1^4 + h_2^4)$$

for a certain constant K independent of h_1 and h_2.
Therefore the global error z satisfies the inequality

$$||z|| \leq K(h_1^4 + h_2^4).$$

This means that

$$||z|| = ||v - u|| = \sqrt{\frac{l_1 l_2}{N_1 N_2}\sum_{i=1}^{N_1-1}\sum_{j=1}^{N_2-1}[v_{ij} - u_{ij}]^2} \to 0,$$

as fast as $h_1^4 + h_2^4 \to 0$ when $N_1, N_2 \to \infty$.

4.2.2   Uniform Convergence in Maximum Norm

We shall use the discrete maximum principle to prove uniform convergence of the finite difference schemes (4.7) and (4.12).
Let us first show that operator $-\Lambda = -\Lambda_x - \Lambda_y$ satisfies the conditions of

definition (1.2), so that $-\Lambda$ is an operator of positive type. Indeed, $-\Lambda$ is a linear operator of the form

$$-\Lambda v(P) = a(P,P)v(P) + \sum_{Q \in N(P)} a(P,Q)v(Q),$$

for all points $P = (ih_1, jh_2) \in \Omega_h$,
The neighbourhood

$$N(P) = \{((i-1)h_1, jh_2), ((i+1)h_1, jh_2), (ih_1, (j-1)h_2), (ih_1, (j+1)h_2)\}$$

Hence, the coefficients of $-\Lambda$ are given by

$$a(P,Q) = \begin{cases} \dfrac{2}{h_1^2} + \dfrac{2}{h_2^2} & if \quad Q = P = (ih_1, jh_2), \\[2mm] -\dfrac{1}{h_1^2} & if \quad Q = ((i-1)h_1, jh_2) \ \ or \ \ Q = ((i+1)h_1, jh_2), \\[2mm] -\dfrac{1}{h_2^2} & if \quad Q = ((ih_1, (j-1)h_2) \ \ or \ \ Q = (ih_1, (j+1)h_2) \end{cases}$$

We note that

$$a(P,Q) < 0, \quad P \neq Q \in N(P), \quad and \quad a(P,P) + \sum_{Q \in N(P)} a(P,Q) \geq 0, \quad P \in \Omega_h,$$

For the points $P \in \Omega_h$ neighbouring the boundary, we have the strong inequality

$$a(P,P) + \sum_{Q \in N(P)} a(P,Q) > 0.$$

Thus, the conditions of definition (3.2) are satisfied, and therefore the operator $-\Lambda_h$ is of positive type. Also, the operator $-L_h = -\Lambda_x - \Lambda_y + c(P)I$ is of positive type, provided that the coefficient $c(P) \geq 0$, for $P \in \Omega_h$, where I is the identity operator.
We shall now state and prove the following:

Theorem 4.1 *The finite difference scheme*

$$-\Lambda_x v_{ij} - \Lambda_y v_{ij} + c_{ij} v_{ij} = f_{ij}, \quad (ih_1, jh_2) \in \Omega_h,$$

$$v_{ij} = \phi_{ij}, \qquad\qquad\qquad (ih_1, jh_2) \in \partial\Omega_h$$

is stable in the maximum norm and the solution v satisfies the following inequality:

$$||v||_{\overline{\Omega}_h} \leq ||\phi||_{\partial\Omega_h} + M \, ||f||_{\Omega_h}, \qquad\qquad (4.28)$$

for $c_{ij} \geq 0$, where M is a constant independent of $h = (h_1, h_2)$.

Proof. In order to prove this theorem, we apply theorem (3.1). Let us choose

$$\beta_h(ih_1, jh_2) = \exp(l_1) - (1 + h_1)^i, \quad i = 0, 1, ..., N_1, \quad j = 0, 1, ..., N_2.$$

as the auxiliary function that appears in the assumption of theorem (3.1). Obviously,

$$0 \le \beta_h(ih_1, jh_2) = \exp(l_1) - (1 + h_1)^i \le \exp(l_1),$$

for $i = 0, 1, ..., N_1, \quad j = 0, 1, ..., N_2$.
Also, we have

$$
\begin{aligned}
L_h[\beta_h](ih_1, jh_2) &= -\Lambda_x \beta_h(ih_1, jh_2) + c_{ij}\beta_h(ih_1, jh_2) \\
&\ge -\frac{1}{h_1^2}[\beta_h((i-1)h_1, jh_2) \\
&\quad -2\beta_h(ih_1, jh_2) + \beta_h((i+1)h_1, jh_2)] \\
&\ge -\frac{1}{h_1^2}[\exp(l_1) - (1+h_1)^{i-1} - 2(\exp(l_1) - (1+h_1)^i \\
&\quad +\exp(l_1) - (1+h_1)^{i+1}] \\
&= \frac{1}{h_1^2}(1+h_1)^{i-1}[1 - 2(1+h_1) + (1+h_1)^2] \\
&\ge (1+h_1)^{i-1} \ge 1,
\end{aligned}
$$

for $i = 1, 2, ..., N_1; \quad j = 1, 2, ..., N_2$.
Thus, assumptions (a) and (b) of theorem (3.1) are satisfied, and by the theorem, inequality (4.28) is true.
Let us now apply the discrete maximum principle to show uniform convergence of the finite difference scheme (4.12).
The operator

$$-L_h = -\Lambda_x - \Lambda_y - \frac{h_1^2 + h_2^2}{12}\Lambda_x\Lambda_y$$

is a positive type, if the steps h_1 and h_2 satisfy the inequalities

$$\frac{1}{\sqrt{5}} \le \frac{h_1}{h_2} \le \sqrt{5}. \tag{4.29}$$

Indeed, $-L_h$ is a linear operator which can be written in the following form:

$$
\begin{aligned}
-L_h v_{ij} =\ & (\frac{2}{h_1^2} + \frac{2}{h_2^2} - \frac{4}{h_1^2 h_2^2}\frac{h_1^2 + h_2^2}{12})v_{ij} \\
& -(\frac{1}{h_1^2} - \frac{2}{h_1^2 h_2^2}\frac{h_1^2 + h_2^2}{12})v_{i+1j} - (\frac{1}{h_2^2} - \frac{2}{h_1^2 h_2^2}\frac{h_1^2 + h_2^2}{12})v_{ij+1} \\
& -(\frac{1}{h_1^2} - \frac{2}{h_1^2 h_2^2}\frac{h_1^2 + h_2^2}{12})v_{i-1j} - (\frac{1}{h_2^2} - \frac{2}{h_1^2 h_2^2}\frac{h_1^2 + h_2^2}{12})v_{ij-1} \\
& -\frac{h_1^2 + h_2^2}{12 h_1^2 h_2^2} v_{i+1j+1} - \frac{h_1^2 + h_2^2}{12 h_1^2 h_2^2} v_{i+1j-1} \\
& -\frac{h_1^2 + h_2^2}{12 h_1^2 h_2^2} v_{i-1j+1} - \frac{h_1^2 + h_2^2}{12 h_1^2 h_2^2} v_{i-1j-1}.
\end{aligned}
$$

The operator $-L_h$ is of positive type if the following inequalities hold (see definition (3.1)):

$$
\frac{1}{h_1^2} - \frac{2}{h_1^2 h_2^2}\frac{h_1^2 + h_2^2}{12} \geq 0 \quad \text{and} \quad \frac{1}{h_2^2} - \frac{2}{h_1^2 h_2^2}\frac{h_1^2 + h_2^2}{12} \geq 0.
$$

Hence, we obtain the inequality

$$
\frac{1}{\sqrt{5}} \leq \frac{h_1}{h_2} \leq \sqrt{5}. \tag{4.30}
$$

Now, from the theorem (3.1), for the auxiliary function $\beta(ih_1, jh_2)$, we conclude that the finite difference scheme (4.12) is uniformly convergent under condition (4.30), and then the global error of the method $z = v - u$ satisfies the inequality

$$
||z|| \leq K(h_1^4 + h_2^4),
$$

where K is a constant independent of h_1 and h_2.

4.3   The Method of Lines for Poisson Equation

Let us consider Poisson equation

$$
\frac{\partial^2 u(x,y)}{\partial^2 x} + \frac{\partial^2 u(x,y)}{\partial y^2} = f(x,y), \quad (x,y) \in R, \tag{4.31}
$$

with the Dirichlet condition

$$
u(x,y) = \begin{cases} \phi_1(y), & x = 0, \\ \phi_2(y), & x = l_1, \\ \psi_1(x), & y = 0, \\ \psi_2(x), & y = l_2. \end{cases}
$$

We shall solve this equation by the method of line. Substituting into (4.31)

$$\frac{\partial^2 u(x, y_i)}{\partial y^2} = \Lambda u(x, y_i) - \frac{h^2}{12} \frac{\partial^4 u(x, \xi_i)}{\partial y^4}, \quad 0 \le x \le l_1,$$

we obtain the semi discrete scheme

$$\frac{\partial^2 u(x, y_i)}{\partial^2 x} + \Lambda u(x, y_i) = f(x, y_i) + E_i(x, h), \quad i = 1, 2, ..., N - 1,$$

where

$$\Lambda u_i(x) = \frac{1}{h^2}[u_{i-1}(x) - 2u_i(x) + u_{i+1}(x)],$$

$$u_i(x) = u(x, y_i), \quad y_i = ih, \ Nh = l_2,$$

and the local truncation error

$$E_i(x, h) = \frac{h^2}{12} \frac{\partial^4 u(x, \xi_i)}{\partial y^4}, \quad \xi_i \in (y_{i-1}, y_{i+1}).$$

Canceling the truncation error in the semi discrete scheme, we get the following system of ordinary differential equations

$$v_i''(x) + \Lambda v_i(x) = f_i(x), \quad i = 1, 2, ..., N - 1, \ 0 \le x \le l_1, \tag{4.32}$$

with the boundary value conditions

$$v_i(0) = \phi_1(ih), \quad v_i(l_1) = \phi_2(ih), \quad i = 1, 2, ..., N - 1,$$

while

$$v_0(x) = \psi_1(x), \quad v_N(x) = \psi_2(x), \quad 0 \le x \le l_1.$$

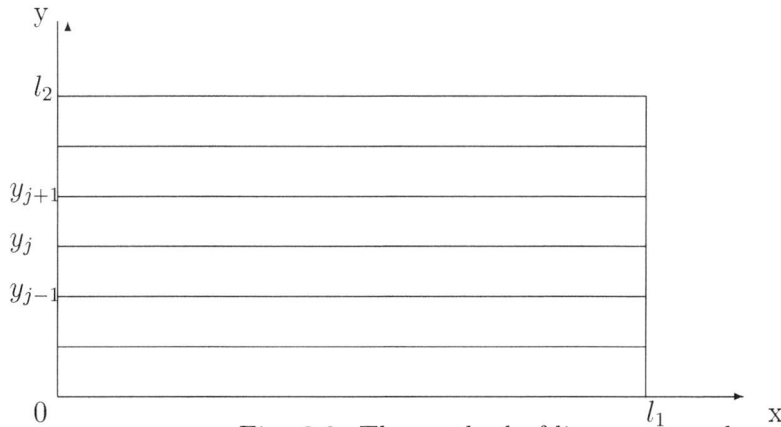

Fig. 3.2. The method of lines - network,

Error estimate. We shall give an estimate of the error of the method

$$z_i(x) = v_i(x) - u_i(x), \quad i = 1, 2, ..., N-1, \ 0 \le x \le l_1,$$

in the maximum norm

$$||z||_{max} = \max_{0 \le x \le l_1} \sqrt{z_1^2(x) + z_2^2(x) + ... + z_{N-1}^2(x)}.$$

It is easy to check that the error $z_i(x)$ satisfies the system of equations

$$z_i''(x) + \Lambda z_i(x) = E_i(x, h), \quad i = 1, 2, ..., N-1, \ \ 0 \le x \le l_1,$$

with the homogeneous boundary conditions

$$z_i(0) = 0, \quad z_i(l_1) = 0, \quad i = 1, 2, ..., N-1,$$

and with

$$z_0(x) = 0, \quad z_N(x) = 0, \quad 0 \le x \le l_1.$$

Let us rewrite this system of equations in the matrix form

$$Z''(x) + M_0 Z = E(x, h), \quad 0 \le x \le l_1,$$

where $Z = [z_1, z_2, ..., z_{N-1}]^T$, $E = [E_1, E_2, ..., E_{N-1}]^T$, and

$$M_0 = \frac{1}{h^2}
\begin{bmatrix}
-2 & 1 & 0 & \cdots & 0 & 0 \\
1 & -2 & 1 & \cdots & 0 & 0 \\
0 & 1 & -2 & \cdots & 0 & 0 \\
\cdots & \cdots & \cdots & \cdots & \cdots & \cdots \\
0 & 0 & 0 & \cdots & -2 & 1 \\
0 & 0 & 0 & \cdots & 1 & -2
\end{bmatrix}.$$

The tri-diagonal matrix M_0 is negative definite, that is

$$(M_0 \mu, \mu) \le -\frac{8}{l_2^2}(\mu, \mu),$$

for all real $\mu = (\mu_1, \mu_2, ..., \mu_{N-1})$.
By the maximum principle (cf. [29]), we have

$$||Z||_{max} \le ||E||_{max} \le \frac{h^2}{12} M^{(4)}, \qquad M^{(4)} = \sup_{x,y} |\frac{\partial^4 u(x, y)}{\partial y^4}|.$$

Thus, the method of lines is convergent and $v_i(x) \to u(x, y_i)$ as fast as $h^2 \to 0$.

4.3.1   Solution of homogeneous equations

Let us consider the homogeneous equations

$$v_i^{''}(x) + \Lambda v_i(x) = 0, \qquad i = 1, 2, ..., N-1, \quad 0 \le x \le l_1, \tag{4.33}$$

with the boundary conditions

$$v_i(0) = \phi_1(ih), \qquad v_i(l_1) = \phi_2(ih), \qquad i = 1, 2, ..., N-1,$$

while

$$v_0(x) = \psi(x) = 0, \qquad v_N(x) = \psi_2(x) = 0 \quad 0 \le x \le l_1,$$

$$\phi_1(0) = \phi_1(l_2) = \phi_2(l_1) = \phi_2(l_2) = 0.$$

Following the method of separation of variables, we substitute into (4.33)

$$v_i(x) = V_i Y(x).$$

Then, we have

$$\frac{Y^{''}(x)}{Y(x)} = -\frac{\Lambda V_i}{V_i} = \sigma^2.$$

Hence

$$-\Lambda V_i = \sigma^2 V_i, \qquad i = 1, 2, ..., N-1,$$

$$V_0 = 0, \qquad V_N = 0, \tag{4.34}$$

and

$$Y^{''}(x) = \sigma^2 Y(x), \qquad 0 \le x \le l_1. \tag{4.35}$$

As we know, the solution of the eigenvalue problem (4.34) is

$$\sigma_k^2 = \frac{4}{h^2} \sin^2 \frac{k\pi h}{2l_2},$$

$$v_i^{(k)} = \sqrt{\frac{2}{l_2}} \sin \frac{k\pi i h}{l_2}, \qquad i, k = 1, 2, ..., N-1.$$

Also, we find the solution of (4.35)

$$Y_k(x) = A_k e^{\sigma_k x} + B_k e^{-\sigma_k x},$$

for arbitrary constants A_k, B_k, $k = 1, 2, ..., N-1$.
Finally, we find the general solution of the homogeneous equations

$$v_i(x) = \sqrt{\frac{2}{l_2}} \sum_{k=1}^{N-1} \sin \frac{k\pi i h}{l_2} [A_k e^{\sigma_k x} + B_k e^{-\sigma_k x}], \qquad i = 0, 1, ..., N, \quad 0 \le x \le l_1.$$

The coefficients A_k and B_k, $k = 1, 2, ..., N - 1$, are determined by the boundary conditions

$$v_i(0) = \phi_1(ih), \quad v_i(l_2) = \phi_2(ih), \quad i = 1, 2, ..., N - 1.$$

Indeed, we have the Fourier sine series representation of $\phi_1(ih)$ and $\phi_2(ih)$, that is

$$\phi_1(ih) = \sqrt{\frac{2}{l_2}} \sum_{k=1}^{N-1} C_k^{(1)} \sin \frac{k\pi ih}{l_2},$$

$$\phi_2(ih) = \sqrt{\frac{2}{l_2}} \sum_{k=1}^{N-1} C_k^{(2)} \sin \frac{k\pi ih}{l_2},$$

where the coefficients

$$C_k^{(1)} = h \sqrt{\frac{2}{l_2}} \sum_{i=1}^{N-1} \phi_1(ih) \sin \frac{k\pi ih}{l_2}, \quad C_k^{(2)} = h \sqrt{\frac{2}{l_2}} \sum_{i=1}^{N-1} \phi_2(ih) \sin \frac{k\pi ih}{l_2}.$$

Comparing the Fourier's coefficients, we get the following equations:

$$A_k + B_k = C_k^{(1)},$$

$$A_k e^{\sigma_k l_1} + B_k e^{-\sigma_k l_1} = C_k^{(2)}, \quad k = 1, 2, ..., N - 1.$$

Hence, we find

$$A_k = \frac{-C_k^{(2)} + C_k^{(1)} e^{-\sigma_k l_1}}{e^{-\sigma_k l_1} - e^{\sigma l_1}}, \quad B_k = \frac{C_k^{(2)} - C_k^{(1)} e^{\sigma_k l_1}}{e^{-\sigma_k l_1} - e^{\sigma l_1}}.$$

Thus, the solution of the homogeneous equation is

$$v_i(x) = \sqrt{\frac{2}{l_2}} \sum_{k=1}^{N-1} \frac{C_k^{(1)} \sinh \sigma_k(l_1 - x) + C_k^{(2)} \sinh \sigma_k x}{\sinh \sigma_k l_1} \sin \frac{k\pi ih}{l_2},$$

for $i = 0, 1, ..., N$, $0 \le x \le l_1$.
Let us note that solution of non-homogeneous equations can be found using the method of variation of parameters.

4.4   Exercises

Question 4.1 *Let $u(x, y)$ be a function six times continuously differentiable in the domain $\Omega = [0, 1] \times [0, 1]$. Then, the finite difference operator*

$$L_h = L_{hx} + L_{hy},$$

where

$$L_{hx} u_{ij} = \frac{1}{12h^2} [-u_{i-2j} + 16u_{i-1j} + 30u_{ij} + 16u_{i+1j} - u_{i+2j}],$$

$$L_{hy} u_{ij} = \frac{1}{12h^2} [-u_{ij-2} + 16u_{ij-1} + 30u_{ij} + 16u_{ij+1} - u_{ij+2}],$$

approximates the Laplace's operator,

$$\Delta u_{ij} = \frac{\partial^2 u_{ij}}{\partial x^2} + \frac{\partial^2 u_{ij}}{\partial y^2},$$

where $u_{ij} = u(ih, jh)$, $i, j = 2, 3, ..., N - 2$, $Nh = 1$.
Determine the truncation error $\psi_{ij}(h) = \Delta u_{ij} - L_h u_{ij}$, $i, j = 2, 3, ..., N - 2$.

Question 4.2 *Solve the following eigenvalue problem:*

$$-\Lambda_x v_{ij} - \Lambda_y v_{ij} + 2v_{ij} = \lambda v_{ij}, \quad (ih, h) \in \Omega_h,$$

$$v_{ij} = 0 \qquad\qquad\qquad (ih, jh) \in \partial\Omega_h$$

where the network $\Omega_h = \{(ih, jh) : i, j = 1, 2, ..., N - 1\}$, $Nh = 2$,
$\partial\Omega_h$ is the boundary of Ω_h.

Question 4.3 *Show that the finite difference operator*

$$L_h v_{ij} = -2\Lambda_x v_{ij} - 3\Lambda_y v_{ij} + 5v_{ij}, \quad (ih_{,h}, jh_2) \in \Omega_h,$$

is positive definite in the Hilbert space $H_h^0(\Omega_h)$ of all discrete functions v_{ij}, $i = 0, 1, ..., N_1$, $j = 0, 1, ..., N_2$, which vanish at the boundary $\partial\Omega_h$ of the network

$$\Omega_h = \{(ih_1, jh_2) : \quad i = 1, 2, ..., N_1 - 1, \quad j = 1, 2, ..., N_2, \quad N_1 h_1 = 2, \quad N_2 h_2 = 4\}.$$

Find a positive lower bound of the eigenvalues λ_{kl}, $k = 1, 2, ..., N_1 - 1$, $l = 1, 2, ..., N_2 - 1$ of the operator L_h.

Question 4.4 *Consider the following boundary value problem:*

$$\frac{\partial^2 u}{\partial x^2} + \frac{\partial^2 u}{\partial y^2} - 5u = \frac{xy}{1 + x^2 + y^2}, \quad \text{in } \Omega = \{(x, y) : 0 < x, y < 1\},$$

$$u(x, y) = \begin{cases} x(1 - x) & \text{if} \quad y = 0, 1, \\ y(1 - y) & \text{if} \quad x = 0, 1. \end{cases}$$

1. (a) *Approximate the boundary problem by an $O(h_1^2 + h_2^2)$ finite difference scheme and determine its truncation error.*

   (b) *Estimate the error of the method in the norm of the Hilbert space $H_h^0(\Omega_h)$ of all discrete functions that vanish at the boundary $\partial\Omega$ of the network Ω_h.*

Question 4.5 *Show that the finite difference operator*

$$L_h = -2\Lambda_x - 4\Lambda_y + 5I,$$

satisfies the maximum principle, where I is the identity operator, and

$$\Lambda_x v_{ij} = \frac{1}{h^2}[v_{i+1j} - 2v_{ij} + v_{i-1j}], \quad \Lambda_y = \frac{1}{h^2}[v_{ij+1} - 2v_{ij} + v_{ij-1}],$$

for $(ih, jh) \in \Omega_h = \{(ih, jh) : i = 1, 2, ..., N-1, \ Nh = 3\}$.

Question 4.6 *Estimate the solution v_{ij}, $i, j = 0, 1, ..., N$, $Nh = 3$ of the finite difference scheme*

$$-\Lambda_x v_{ij} - 2\Lambda_y v_{ij} + 3v_{ij} = (i + j)h, \qquad (ih, jh) \in \Omega_h,$$

$$v_{ij} = 4ijh^2 \qquad\qquad\qquad (ih, jh) \in \partial\Omega_h,$$

using the maximum principle.

Question 4.7 *Consider the following boundary value problem:*

$$2\frac{\partial^2 u}{\partial x^2} + 3\frac{\partial^2 u}{\partial y^2} - 4u = \exp(-x^2 - y^2), \quad \text{in } \Omega = \{(x, y) : \ 0 < x, y < 1\},$$

$$u(x, y) = \begin{cases} \sin \pi x & if \ \ y = 0, 1, \\ \sin \pi y & if \ \ x = 0, 1. \end{cases}$$

1. (a) *Approximate the boundary problem by an $O(h^2)$ finite difference scheme and determine its truncation error.*

   (b) *Estimate the error of the method in the maximum norm.*

Question 4.8 *Show that the nonlinear elliptic equation*

$$\frac{\partial^2 u}{\partial x^2} + \frac{\partial^2 u}{\partial y^2} + \exp(-\frac{\partial^2 u}{\partial x^2} - \frac{\partial^2 u}{\partial y^2}) = 1 \quad \text{in } \Omega = \{(x, y) : \ 0 < x, y < 1\},$$

$$(4.36)$$

satisfies the assumptions of the maximum principle.
Approximate the equation (4.36) along with the Dirichlet's boundary value condition

$$u(x, y) = \sin \pi x \sin \pi y, \quad (x, y) \in \partial\Omega,$$

by an $O(h^2)$ uniformly convergent finite difference scheme. Estimate the error of the method in the maximum norm.

Question 4.9 *Use the method of lines to solve the equation*

$$\frac{\partial^2 u(x,y)}{\partial^2 x} + \frac{\partial^2 u(x,y)}{\partial y^2} = \sin 2\pi x, \quad 0 \le x, y \le 1,$$

with the boundary condition

$$u(x,y) = \begin{cases} \sin 3\pi y, & x = 0, \\ \sin 5\pi y, & x = 1, \\ 0, & y = 0, \\ 0, & y = 1. \end{cases}$$

Give an estimate of the error of the method in the maximum norm.

Chapter 5

Parabolic Equations

Tadeusz Styš
University of Warsaw

Abstract: In this chapter, considerable attention is paid to construction of finite difference schemes with weight $0 \leq \sigma \leq 1$ for diffusion equations in one and two space variables. Efficient algorithms are built for selected values of the weight σ, like implicit scheme ($\sigma = 1$), explicit scheme ($\sigma = 0$) and Crank Nicolson scheme ($\sigma = 1/2$). The average convergence of the scheme is proved in the norm of the Hilbert's space H. The scheme with weight is implemented in the *Mathematica* module *heatEqn* and applied to the diffusion equations with initial boundary value conditions. In the last section, the heat equation with initial boundary value conditions is solved by the method of lines. The chapter ends with a set of questions.

Chapter 5

Parabolic Equations

5.1　Introduction

Parabolic equations have been used in various areas of sciences for modeling non-stationary phenomena. In general, these equations may be considered in the following form:

$$\frac{\partial u}{\partial t} = L(u),$$

where $L(u)$ is an elliptic operator.

The simplest example of a parabolic equation is the heat equation

$$\frac{\partial u(t,x)}{\partial t} = a^2 \frac{\partial^2 u(t,x)}{\partial x^2} + f(t,x), \quad t \geq 0, \quad 0 \leq x \leq l, \qquad (5.1)$$

with the initial condition

$$u(0, x = \phi(x), \quad 0 \leq x \leq l, \qquad (5.2)$$

and with the boundary condition

$$u(t, 0) = \nu_0(t), \qquad u(t, l) = \nu_1(t), \quad t \geq 0, \qquad (5.3)$$

where $f(t,x)$, $\phi(x)$, $\nu_0(t)$ and $\nu_1(t)$ are given continuous functions for $t \geq 0$ and $0 \leq x \leq l$.

We shall use the following notation:

$$t_n = n\tau, \quad n = 0, 1, ..., \frac{T}{\tau}, \quad x_i = ih, \quad i = 0, 1, ..., N, \quad Nh = l,$$

$$\Omega_h = \{ih \ : \ i = 1, 2, ..., N-1\}, \quad u_i^n = u(n\tau, ih).$$

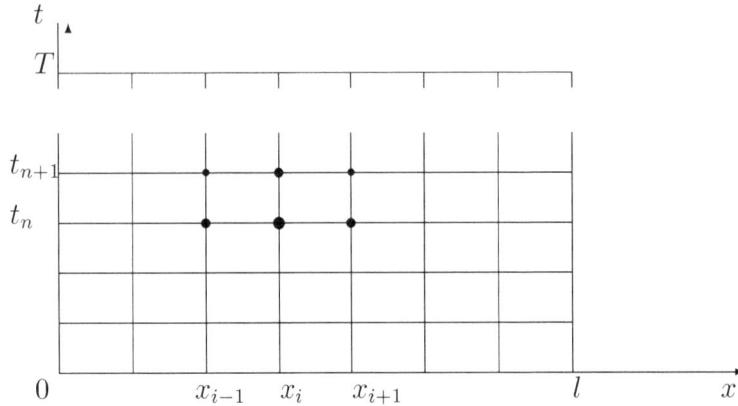

Fig 5.1. Network Ω_h

The family of finite difference schemes with a weight includes well established methods for solving diffusion equations. We shall study these schemes in the next sections.

5.2   The Finite Difference Scheme with Weight for the Heat Equation in One Space Variable

Let us write the heat equation (5.1) in the following form:

$$\frac{\partial u(t,x)}{\partial t} = a^2 [\sigma \frac{\partial^2 u(t,x)}{\partial x^2} + (1-\sigma)\frac{\partial^2 u(t,x)}{\partial x^2}] + f(t,x), \qquad (5.4)$$

where the weight $0 \le \sigma \le 1$.

If $u(t,x)$ is the solution of (5.4) at least twice continuously differentiable with respect to variable t and four times continuously differentiable with respect to variable x, then

$$\frac{u_i^{n+1} - u_i^n}{\tau} = \frac{\partial u_i^n}{\partial t} + \frac{\tau}{2}\frac{\partial^2 u(\zeta_n, ih)}{\partial t}, \qquad \zeta_n \in (t_n, t_{n+1}),$$

$$\Lambda u_i^n \equiv \frac{1}{h^2}(u_{i-1}^n - 2u_i^n + u_{i+1}^n)$$

$$= \frac{\partial^2 u_i^n}{\partial x^2} + \frac{h^2}{12}\frac{\partial^4 u(n\tau, \xi_i)}{\partial x^4}, \qquad \xi_i \in (x_{i-1}, x_{i+1}), \qquad (5.5)$$

$$\Lambda u_i^{n+1} \equiv \frac{1}{h^2}(u_{i-1}^{n+1} - 2u_i^{n+1} + u_{i+1}^{n+1})$$

$$= \frac{\partial^2 u_i^{n+1}}{\partial x^2} + \frac{h^2}{12}\frac{\partial^4 u((n+1)\tau, \xi_i)}{\partial x^4}, \quad \xi_i \in (x_{i-1}, x_{i+1}),$$

Using (5.4) and (5.5), it is easy to show that the theoretical solution satisfies the following equation:

$$\frac{u_i^{n+1} - u_i^n}{\tau} = a^2 \Lambda [\sigma u_i^{n+1} + (1-\sigma)u_i^n] + \Theta_i^n + \psi_i^n(\tau, h), \qquad (5.6)$$

at the mesh points $(n\tau, ih)$, $n = 0, 1, ..., \frac{T}{\tau}$, $i = 1, 2, ..., N-1$.
Here, the term Θ_i^n depends on $f(t, x)$, and the local truncation error

$$\psi_i^n(\tau, h) = \frac{u_i^{n+1} - u_i^n}{\tau} - a^2 \Lambda [\sigma u_i^{n+1} + (1-\sigma)u_i^n] - \Theta_i^n. \qquad (5.7)$$

Cancelling the local truncation error in equation (5.6), we arrive at the following finite difference scheme with the weight σ (cf. [25])

$$\frac{v_i^{n+1} - v_i^n}{\tau} = a^2 \Lambda [\sigma v_i^{n+1} + (1-\sigma)v_i^n] + \Theta_i^n, \quad i = 1, 2, ..., N-1,,$$

$$v_0^n = \nu_0^n, \qquad v_N^n = \nu_1^n, \qquad\qquad\qquad n = 0, 1, ..., \frac{T}{\tau}. \qquad (5.8)$$

$$v_i^0 = \phi_i, \qquad\qquad\qquad\qquad\qquad i = 0, 1, ..., N.$$

5.2.1   Estimate of the Local Truncation Error

We shall express the error $\psi_i^n(\tau, h)$ in terms of the mesh steps τ and h. Namely, using Taylor's formula at the point $((n+0.5)\tau, ih)$, we have

$$u_i^n = u_i^{n+0.5} - \frac{\tau}{2}\frac{\partial u_i^{n+0.5}}{\partial t} + \frac{\tau^2}{8}\frac{\partial^2 u_i^{n+0.5}}{\partial t^2} - \frac{\tau^3}{48}\frac{\partial^3 u(\zeta_n, ih)}{\partial t^3}, \qquad \zeta_n \in (t_n, t_{n+1}),$$

$$u_i^{n+1} = u_i^{n+0.5} + \frac{\tau}{2}\frac{\partial u_i^{n+0.5}}{\partial t} + \frac{\tau^2}{8}\frac{\partial^2 u_i^{n+0.5}}{\partial t^2} + \frac{\tau^3}{48}\frac{\partial^3 u(\zeta_n, ih)}{\partial t^3}, \quad \zeta_n \in (t_n, t_{n+1}),$$

$$\sigma u_i^{n+1} + (1-\sigma)u_i^n = u_i^{n+0.5} + (2\sigma - 1)\frac{\tau}{2}\frac{\partial u_i^{n+0.5}}{\partial t} + \frac{\tau^2}{8}\frac{\partial^2 u_i^{n+0.5}}{\partial t^2} + O(\tau^3),$$

and

$$a^2 \Lambda [\sigma u_i^{n+1} + (1-\sigma)u_i^n] = a^2 \frac{\partial^2 u_i^{n+0.5}}{\partial x^2} + \frac{a^2 h^2}{12}\frac{\partial^4 u_i^{n+0.5}}{\partial x^4}$$

$$+ (2\sigma - 1)\frac{a^2 \tau}{2}\Lambda\frac{\partial u_i^{n+0.5}}{\partial t} + O(\tau^2 + h^4),$$

$$\Lambda\frac{\partial u_i^{n+0.5}}{\partial t} = a^2 \frac{\partial^4 u_i^{n+0.5}}{\partial x^4} + \frac{a^2 h^2}{12}\frac{\partial^6 u(\zeta_n, ih)}{\partial x^6} + \Lambda f_i^{n+0.5} + O(h^4).$$

Hence, by (5.7)

$$
\psi_i^n(\tau,h) = \ [\frac{\partial u_i^{n+0.5}}{\partial t} - a^2 \frac{\partial^2 u_i^{n+0.5}}{\partial x^2} - \Theta_i^n] - [(2\sigma-1)\frac{a^4\tau}{2} + \frac{a^2 h^2}{12}]\frac{\partial^4 u_i^{n+0.5}}{\partial x^4}
$$
$$
- (2\sigma-1)\frac{a^2\tau h^2}{24}[\frac{\partial^6 u(\zeta_n,ih)}{\partial x^6} + \frac{\partial^4 f_i^{n+0.5}}{\partial x^4}] + O(\tau^2 + h^4).
$$

$$(5.9)$$

Finally, the local truncation error

$$
\psi_i^n(\tau,h) = \begin{cases}
O(\tau + h^2) & if \ \sigma \neq \frac{1}{2}, \quad \sigma \neq \frac{1}{2} - \frac{h^2}{12\tau a^2}, \quad \Theta_i^n = f_i^n, \quad u \in C_4^{(2)}, \\[2mm]
O(\tau^2 + h^2) & if \ \sigma = \frac{1}{2}, \quad \Theta_i^n = f_i^{n+0.5}, \qquad\qquad\qquad u \in C_4^{(3)}, \\[2mm]
O(\tau^2 + h^4) & if \ \sigma = \frac{1}{2} - \frac{h^2}{12\tau a^2}, \quad \Theta_i^n = f_i^{n+0.5}, \qquad u \in C_6^{(3)},
\end{cases}
$$

$$(5.10)$$

where $C_q^{(k)}$ is the class of functions which are k-times continuously differentiable with respect to variable t and q times continuously differentiable with respect to variable x.

Now, let us select from the class (5.8) the standard finite difference schemes.

5.2.2   The Explicit Finite Difference Scheme

Setting $\sigma = 0$ and $\Theta_i^n = f_i^n$ in (5.8), we obtain the following *explicit finite difference scheme*

$$
\frac{v_i^{n+1} - v_i^n}{\tau} = a^2 \Lambda v_i^n + f_i^n, \quad i = 1,2,...,N-1, \quad n = 0,1,...,\frac{T}{\tau},
$$
$$
v_i^0 = \phi_i, \qquad\qquad\qquad i = 0,1,...,N,
$$
$$
v_0^n = \nu_0^n, \quad v_N^n = \nu_1^n, \qquad n = 0,1,...\frac{T}{\tau},
$$

$$(5.11)$$

which has the local truncation error $\psi_i^n = O(\tau + h^2)$.

Clearly, the explicit finite difference scheme can be written in the following form:

$$
v_i^{n+1} = v_i^n + a^2\tau \Lambda v_i^n + \tau f_i^n, \quad i = 1,2,...,N-1, \quad n = 0,1,...,\frac{T}{\tau}.
$$

5.2.3   The Pure Implicit Finite Difference Scheme

Setting $\sigma = 1$ and $\Theta_i^n = f_i^{n+1}$ in (5.8), we obtain the following *pure implicit finite difference scheme*

$$\frac{v_i^{n+1} - v_i^n}{\tau} = a^2 \Lambda v_i^{n+1} + f_i^{n+1}, \quad i = 1, 2, ..., N-1, \quad n = 0, 1, ..., \frac{T}{\tau},$$

$$v_i^0 = \phi_i, \qquad\qquad\qquad\quad i = 0, 1, ..., N, \qquad\qquad (5.12)$$

$$v_0^n = \nu_0^n, \quad v_N^n = \nu_1^n, \qquad\quad n = 0, 1, ...\frac{T}{\tau},$$

which has the local truncation error $\psi_i^n = O(\tau + h^2)$.

Let us note that the implicit finite difference scheme (5.12), is the following system of $N-1$ linear equations with the tri-diagonal matrix

$$\begin{bmatrix} 1 + \dfrac{2\tau a^2}{h^2} & -\dfrac{\tau a^2}{h^2} & \cdots & 0 \\[2mm] -\dfrac{\tau a^2}{h^2} & 1 + \dfrac{2\tau a^2}{h^2} & \cdots & 0 \\[2mm] \vdots & \vdots & \ddots & \vdots \\[2mm] 0 & 0 & \cdots & 1 + \dfrac{2\tau a^2}{h^2} \end{bmatrix} \begin{bmatrix} v_1^{n+1} \\[2mm] v_2^{n+1} \\[2mm] \vdots \\[2mm] v_{N-1}^{n+1} \end{bmatrix} = \begin{bmatrix} v_1^n + \tau f_1^{n+1} + \dfrac{\tau a^2}{h^2} v_0^{n+1} \\[2mm] v_2^n + \tau f_2^{n+1} \\[2mm] \vdots \\[2mm] v_{N-1}^n + \tau f_{N-1}^{n+1} + \dfrac{\tau a^2}{h^2} v_N^{n+1} \end{bmatrix}$$

5.2.4   Crank-Nicolson Finite Difference Scheme.

Setting $\sigma = \frac{1}{2}$ and $\Theta_i^n = f_i^{n+0.5}$ in (5.8), we obtain the *Crank-Nicolson finite difference scheme*

$$\frac{v_i^{n+1} - v_i^n}{\tau} = \frac{a^2}{2} \Lambda[v_i^{n+1} + v_i^n] + f_i^{n+0.5} \quad i = 1, 2, ..., N-1, ,$$

$$v_0^n = \nu_0^n, \quad v_N^n = \nu_1^n, \qquad\qquad n = 0, 1, ...\frac{T}{\tau}, \qquad\qquad (5.13)$$

$$v_i^0 = \phi_i, \qquad\qquad\qquad\qquad i = 0, 1, ..., N,$$

which has the local truncation error $\psi_i^n = O(\tau^2 + h^2)$.

5.2.5   The Higher Order Finite Difference Scheme.

Setting $\sigma = \dfrac{1}{2} - \dfrac{h^2}{12\tau a^2}$ and

$$\Theta_i^n = f_i^{n+0.5} + \frac{h^2}{12} \Lambda f_i^{n+0.5} = \frac{1}{12}(f_{i-1}^{n+0.5} + 10 f_i^{n+0.5} + f_i^{n+0.5}),$$

in (5.8), we obtain the *higher order finite difference scheme*

$$\frac{v_i^{n+1} - v_i^n}{\tau} = a^2 \Lambda \left[\frac{1}{2}(v_i^{n+1} + v_i^n) - \frac{h^2}{12\tau a^2}(v_i^{n+1} - v_i^n) \right]$$

$$+ \frac{1}{12}(f_{i-1}^{n+0.5} + 10 f_i^{n+0.5} + f_{i+1}^{n+0.5}), i = 1, 2, ..., N-1, \qquad (5.14)$$

$$v_0^n = \nu_0^n, \quad v_N^n = \nu_1^n, \quad n = 0, 1, ..., \frac{T}{\tau},$$

$$v_i^0 = \phi_i, \quad i = 0, 1, ..., N,$$

This scheme has the local truncation error $\psi_i^n = O(\tau^2 + h^4)$ if $\sigma = \dfrac{1}{2} - \dfrac{h^2}{12\tau a^2}$ and $0 \le \sigma \le 1$.

5.2.6   Stability in the Hilbert Space H_h^0.

Stability of a finite difference scheme can be considered in a specific norm. At this point, we shall deal with the stability in the norm of the Hilbert space H_h^0.

Let us write the finite difference scheme with the weight σ as follows:

$$(I - a^2 \tau \sigma \Lambda) v_i^{n+1} = (I + a^2 \tau (1 - \sigma) \Lambda) v_i^n + \tau \Theta_i^n,$$

$$i = 1, 2, ..., N-1, \quad n = 0, 1, ..., \frac{T}{\tau}, \qquad (5.15)$$

$$v_i^0 = \phi_i, \quad i = 0, 1, ..., N,$$

with the homogeneous boundary condition

$$v_0^n = v_N^n = 0, \quad n = 0, 1, ..., \frac{T}{\tau},$$

where I is the identity operator.

Since the operator $I - a^2 \tau \sigma \Lambda$ is positive definite in the Hilbert space H_h^0 with the lower-bound of its eigenvalues $\gamma = 1 \le 1 + a^2 \tau \sigma \lambda_1$, there exists the inverse operator

$$(I - a^2 \tau \sigma \Lambda)^{-1}$$

and the norm

$$||(I - a^2 \tau \sigma \Lambda)^{-1}|| \le 1.$$

Thus, we can write equation (5.15) in the equivalent form

$$v_i^{n+1} = (I - a^2 \tau \sigma \Lambda)^{-1}(I + a^2 \tau (1 - \sigma) \Lambda) v_i^n + \tau (I - a^2 \tau \sigma \Lambda)^{-1} \Theta_i^n, \qquad (5.16)$$

for $i = 1, 2, ..., N-1, \quad n = 0, 1, ..., \dfrac{T}{\tau}.$

Now, multiplying (5.16) by $v^{n+1} \in H_h^0$ in the sense of the inner product of H_h^0, we obtain

$$(v^{n+1}, v^{n+1}) = (I - a^2\tau\sigma\Lambda)^{-1}((I + a^2\tau(1-\sigma)\Lambda v^n, v^{n+1})$$

$$+ \tau((I - a^2\tau\sigma\Lambda)^{-1}\Theta^n, v^{n+1}),$$

for $n = 0, 1, ..., \frac{T}{\tau}$.

Hence, by the Cauchy inequality

$$||v^{n+1}||^2 \leq ||(I - a^2\tau\sigma\Lambda)^{-1}(I + a^2\tau(1-\sigma)\Lambda)|| \, ||v^n|| \, ||v^{n+1}||$$

$$+ \tau||(I - a^2\tau\sigma\Lambda)^{-1}|| \, ||\Theta^n|| \, ||v^{n+1}||, \tag{5.17}$$

or

$$||v^{n+1}|| \leq ||(I - a^2\tau\sigma\Lambda)^{-1}(I + a^2\tau(1-\sigma)\Lambda)|| \, ||v^n||$$

$$+ \tau||(I - a^2\tau\sigma\Lambda)^{-1}|| \, ||\Theta^n||, \tag{5.18}$$

for $n = 0, 1, ..., \frac{T}{\tau}$.

Let us note that

$$||(I - a^2\tau\sigma\Lambda)^{-1}(I + a^2\tau(1-\sigma)\Lambda)|| = \max_{1 \leq k \leq N-1} |\frac{1 - a^2\tau(1-\sigma)\lambda_k}{1 + a^2\tau\sigma\lambda_k}|,$$

where

$$\lambda_k = \frac{4}{h^2}\sin^2\frac{k\pi h}{2l}, \quad k = 1, 2, ..., N-1,$$

are eigenvalues of the operator $-\Lambda$.

In order to get an upper-bound of $||v^{n+1}||$ via (5.18), the inequality

$$|\frac{1 - a^2\tau(1-\sigma)\lambda_k}{1 + a^2\tau\sigma\lambda_k}| \leq 1$$

must be satisfied for all $k = 1, 2, ..., N-1$.

This means that

$$-1 \leq \frac{1 - a^2\tau(1-\sigma)\lambda_k}{1 + a^2\tau\sigma\lambda_k} \leq 1.$$

Hence

$$\sigma \geq \frac{1}{2} - \frac{1}{a^2\tau\lambda_k}, \quad k = 1, 2, ..., N-1.$$

Since $\lambda_k \leq \frac{4}{h^2}, \quad k = 1, 2, ..., N-1$, therefore the condition of stability is:

$$\sigma \geq \frac{1}{2} - \frac{h^2}{4a^2\tau}. \tag{5.19}$$

Obviously, under condition (5.19) and by the inequality (5.18)

$$||v^n|| \leq ||v^0|| + \tau(||\Theta^0|| + ||\Theta_1|| + \cdots + ||\Theta^{\frac{T}{\tau}}||), \tag{5.20}$$

for all $n = 0, 1, ..., \frac{T}{\tau}$.

Since

$$\tau(||\Theta^0|| + ||\Theta_1|| + \cdots + ||\Theta^{\frac{T}{\tau}}||) \le T \max_{0 \le n \le \frac{T}{\tau}} ||\Theta^n||,$$

we obtain

$$||v^n|| \le ||\phi|| + T \max_{0 \le n \le \frac{T}{\tau}} ||\Theta^n||, \qquad (5.21)$$

for all $n = 0, 1, ..., \frac{T}{\tau}$.

Thus, the finite difference scheme with weight σ is stable in the norm of the Hilbert space H_h^0 provided that the weight σ satisfies condition (5.19).

Now, let us note that the finite difference scheme with weight σ is unconditionally stable if $\sigma \ge \frac{1}{2}$, and it is conditionally stable if $\sigma < \frac{1}{2}$. Therefore

- the pure implicit finite difference scheme (5.12), ($\sigma = 1$) is unconditionally stable,

- Crank-Nicolson's scheme (5.13), ($\sigma = \frac{1}{2}$) is unconditionally stable,

- the higher order finite difference scheme (5.14), ($\sigma = \frac{1}{2} - \frac{h^2}{12\tau a^2}$) is unconditionally stable.

- the explicit finite difference scheme ((5.11), $\sigma = 0$) is conditionally stable, under the condition

$$\frac{\tau}{h^2} \le \frac{1}{2a^2}.$$

5.2.7   Convergence

The finite difference scheme (5.8) is convergent if the approximate solution v_i^n converges to the theoretical solution u_i^n at any fixed mesh point $(n^*\tau, i^*h)$, when $\tau, h \rightarrow 0$.

This means that the global error of the method

$$z_i^n = v_i^n - u_i^n, \quad n = 0, 1, ..., \frac{T}{\tau}, \quad i = 0, 1, ..., N.$$

tends to zero when $, \tau \rightarrow 0$.

Clearly, the error z is the solution of the finite difference scheme

$$\frac{z_i^{n+1} - z_i^n}{\tau} = a^2 \Lambda[\sigma z_i^{n+1} + (1 - \sigma)z_i^n] + \psi_i^n(\tau, h), \quad i = 1, 2, ..., N - 1,$$

$$z_0^n = z_N^N = 0, \qquad\qquad n = 0, 1, ..., \frac{T}{\tau},$$

$$z_i^0 = 0, \qquad\qquad i = 0, 1, ..., N,$$

where $\psi_i^n(\tau, h)$ is the local truncation error given by formula (5.10).
Applying the inequality (5.21)) to this finite difference scheme, we obtain the following error estimate:

$$||z^n|| \leq T \max_{0 \leq n \leq \frac{T}{\tau}} ||\psi^n||$$

for all $n = 0, 1, ..., \frac{T}{\tau}$, provided that the stability condition (5.19) holds.
By formula (5.10), the local truncation error

$$||\psi^n(\tau, h)|| \leq K \begin{cases} \tau + h^2 & if \ \sigma \neq \frac{1}{2}, \ \ \sigma \neq \frac{1}{2} - \frac{h^2}{12\tau a^2}, \ \ u \in C_4^{(2)}, \\ \tau^2 + h^2 & if \ \sigma = \frac{1}{2}, & u \in C_4^{(3)}, \\ \tau^2 + h^4 & if \ \sigma = \frac{1}{2} - \frac{h^2}{12\tau a^2}, & u \in C_6^{(3)}, \end{cases} \quad (5.22)$$

where the generic constant K is independent of τ and h.
Therefore, the finite difference scheme with the weight σ is convergent under condition (5.19). The rate of convergence is determined by the local truncation error. For instance, the explicit finite difference scheme is convergent with the rate $O(\tau + h^2)$ if $\frac{\tau}{h^2} \leq \frac{1}{2}$. The pure implicit finite difference scheme is unconditionally convergent with the rate $O(\tau + h^2)$. The Crank-Nicolson scheme is also unconditionally convergent and its rate of convergence is $O(\tau^2 + h^2)$. The higher order finite difference scheme converges with the rate $O(\tau^2 + h^4)$ provided that condition (5.19) holds.

Example 5.1 *Consider the hear equation*

$$\frac{\partial u(t, x)}{\partial t} = 4\frac{\partial^2 u(t, x)}{\partial x^2} + (4\pi^2 - 1)e^{-t}\sin \pi x, \quad 0 \leq x \leq 1, \ \ t \geq 0,$$

with the initial value condition

$$u(0, x) = \sin \pi x, \quad 0 \leq x \leq 1,$$

and with the homogeneous boundary conditions $u(t, 0) = u(t, 1) = 0$, $t \geq 0$.

Let us solve the heat equation using the following module

Program 5.1 *Mathematica module that solves finite difference scheme with weight for heat equation*

```
heatEqn[f_,phi_,a_,sigma_,nx_,nt_,tau_]:=Module[
        {c,d,f1,laplace,lh,soltri,rozw},
h=1/nx;
d=1+2*a^2 sigma*tau/h^2;   c=-a^2 tau*sigma/h^2;

f1[k_]:=Table[f[k*tau,i*h],{i,1,nx-1}];
v=Table[phi[i*h],{i,0,nx}];

laplace[v_,h_]:=Take[(RotateRight[v,1]-2 v +
                RotateLeft[v,1])/h^2,{2,-2}];

soltri[k_,v_]:=Module[{al,be,x},
        lh=Take[v,{2,-2}]+a^2 tau (1-sigma)*laplace[v,h]+tau*f1[k];
        al[1]=c/d;
        al[i_]:=al[i]=c/(d-al[i-1]*c);
        be[1]=lh[[1]]/d;
        be[i_]:=be[i]=(lh[[i]]-be[i-1]*c)/(d-al[i-1]*c);
        x[nx-1]=be[nx-1];
        x[i_]:=x[i]=be[i]-al[i]*x[i+1];
        Append[Prepend[Table[x[i],{i,1,nx-1}],0],0]
    ];
rozw={ };
    Do[{AppendTo[rozw, v];v=soltri[j,v]},{j,1,nt}];
    rozw
 ]
```

We enter the following input data

```
            a=2;  sigma=1; nx=5; nt=5;
            f[t_,x_]:=(a^2 Pi^2-1)Exp[-t] Sin[Pi x];
            phi[x_]:=N[Sin[Pi x]];
```

where nx and nt determine the number of points on x and t axes.
Executing the command

```
    N[heatEqn[f,phi,2,1,5,5,0.1],3]//TableForm
```

we obtain the following table of the approximate solution $v(n\tau, ih)$

t\x	0	0.2	0.4	0.6	0.8	1
0	0	0.588	0.951	0.951	0.588	0
0.1	0	0.532	0.861	0.861	0.532	0
0.2	0	0.481	0.779	0.779	0.481	0
0.3	0	0.435	0.705	0.705	0.435	0
0.4	0	0.394	0.638	0.638	0.394	0

Comparing the approximate solution $v(n\tau, ih)$ with the theoretical solution $u(n\tau, ih) = e^{n\tau} \sin ih$ at the grid points, we determine the absolute error 0.02648 which corresponds with the estimate of the global error $O(\tau + h^2)$, for $h = 0.2$ and $\tau = 0.1$.

5.3   The Finite Difference Scheme with Weight for the Heat Equation in Two Space Variables.

Let us consider the heat equation

$$\frac{\partial u}{\partial t} = \frac{\partial^2 u}{\partial x^2} + \frac{\partial^2 u}{\partial y^2} + f(t, x, y), \qquad in \ \ [0, \infty) \times \Omega, \qquad (5.23)$$

with the initial value condition

$$u(0, x, y) = \phi(x, y), \qquad on \ \ \Omega, \qquad (5.24)$$

and with the boundary value condition

$$u(t, x, y) = \nu(t, x, y), \qquad on \ \ [0, \infty) \times \partial\Omega, \qquad (5.25)$$

where $\partial\Omega$ is the boundary of the rectangle

$$\Omega = \{(x, y) \ : \ \ 0 \leq x \leq l_1, \ 0 \leq y \leq l_2\}.$$

We shall use the following notation:

$$\overline{\Omega}_h = \{(ih_1, jh_2) \ : \ \ i = 0, 1, ..., N_1; \ j = 0, 1, ..., N_2; \ N_1 h_1 = l_1, \ N_2 h_2 = l_2\},$$

$$\Gamma_\tau = \{(t_n, ih_1, jh_2) \ : \ \ t_n = n\tau, \ \ n = 0, 1, ..., \tfrac{T}{\tau}, \ \ (ih_1, jh_2) \in \partial\Omega\},$$

$$\partial\Omega_h = \{(ih_1, jh_2) \in \Omega\}, \qquad \Omega_h = \overline{\Omega}_h - \partial\Omega_h, \qquad u_{ij}^n = u(n\tau, ih_1, jh_2),$$

$$Q_{h,\tau} = \Gamma_\tau \times \Omega_h,$$

$$\Lambda_x u_{ij}^n = \frac{1}{h_1^2}[u_{i-1j}^n - 2u_{ij}^n + u_{i+1j}^n], \qquad \Lambda_y u_{ij}^n = \frac{1}{h_2^2}[u_{ij-1}^n - 2u_{ij}^n + u_{ij+1}^n].$$

Clearly, we can write equation (5.23) as follows:

$$\frac{\partial u}{\partial t} = \sigma_1 \frac{\partial^2 u}{\partial x^2} + (1 - \sigma_1)\frac{\partial^2 u}{\partial x^2} + \sigma_2 \frac{\partial^2 u}{\partial y^2} + (1 - \sigma_2)\frac{\partial^2 u}{\partial y^2} + f(t, x, y),$$

where the weights $0 \leq \sigma_1, \sigma_2 \leq 1$.
Replacing the derivatives by corresponding finite difference schemes, in (5.23), we obtain the following equation:

$$\frac{u_{ij}^{n+1} - u_{ih}^n}{\tau} = \Lambda_x[\sigma_1 u_{ij}^{n+1} + (1 - \sigma_1)u_{ij}^n] + \Lambda_y[\sigma_2 u_{ij}^{n+1} + (1 - \sigma_2)u_{ij}^n] + \Theta_{ij}^n + \psi_{ij}^n,$$

$$(5.26)$$

at the points $(n\tau, ih_1, jh_2) \in Q_{h,\tau} = \{0, \tau, 2\tau, ..., T\} \times \Omega_h$,
where Θ will be determined later on, ψ is the local truncation error given by
the formula

$$\psi_{ij}^n = \frac{u_{ij}^{n+1} - u_{ij}^n}{\tau} - \Lambda_x[\sigma_1 u_{ij}^{n+1} + (1 - \sigma_1)u_{ij}^n] - \Lambda_y[\sigma_2 u_{ij}^{n+1} + (1 - \sigma_2)u_{ij}^n] - \Theta_{ij}^n,$$
(5.27)

Cancelling the local truncation error in (5.26), we arrive at the finite difference
scheme with weight

$$\frac{v_{ij}^{n+1} - v_{ij}^n}{\tau} = \Lambda_x[\sigma_1 v_{ij}^{n+1} + (1 - \sigma_1)v_{ij}^n] + \Lambda_y[\sigma_2 v_{ij}^{n+1} + (1 - \sigma_2)v_{ij}^n] + \Theta_{ij}^n,$$
(5.28)

for $(n\tau, ih_1, jh_2) \in Q_{h,\tau}$,
with the initial-boundary conditions

$$\begin{aligned} v_{ij}^0 &= \phi_{ij}, && (ih_1, jh_2) \in \Omega_h, \\ v_{ij}^n &= \nu_{ij}^n, && (n\tau, ih_1, jh_2) \in \Gamma_\tau. \end{aligned}$$
(5.29)

As in one space variable, we can select from the family of schemes (5.28) the
well established finite difference schemes listed below.

5.3.1   The Explicit Finite Difference Scheme.

Setting $\sigma_1 = \sigma_2 = 0$ and $\Theta_{ij}^n = f_{ij}^n$, in (5.28), we obtain the *explicit finite
difference scheme*

$$\begin{aligned} \frac{v_{ij}^{n+1} - v_{ij}^n}{\tau} &= \Lambda_x v_{ij}^n + \Lambda_y v_{ij}^n + \Theta_{ij}^n, && (n\tau, ih_1, jh_2) \in \Gamma_\tau \times \Omega_h, \\ v_{ij}^0 &= \phi_{ij}, && (ih_1, jh_2) \in \Omega_h, \\ v_{ij}^n &= \nu_{ij}^n, && (n\tau, ih_1, jh_2) \in \Gamma_\tau. \end{aligned}$$
(5.30)

Hence, we can get the solution v in the explicit form

$$\begin{aligned} v_{ij}^{n+1} &= v_{ij}^n + \tau[\Lambda_x v_{ij}^n + \Lambda_y v_{ij}^n] + f_{ij}^n], && (n\tau, ih_1, jh_2) \in Q_{h,\tau}, \\ v_{ij}^0 &= \phi_{ij}, && (ih_1, jh_2) \in \Omega_h, \\ v_{ij}^n &= \nu_{ij}^n, && (n\tau, ih_1, jh_2) \in \Gamma_\tau. \end{aligned}$$
(5.31)

The explicit finite difference scheme has the local truncation error $\psi_{ij}^n = O(\tau + h_1^2 + h_2^2)$ and it is conditionally convergent.

5.3.2   The Pure Implicit Finite Difference Scheme.

Setting $\sigma_1 = \sigma_2 = 1$ and $\Theta_{ij}^n = f_{ij}^{n+1}$ in (5.28), we obtain the *pure implicit finite difference scheme*

$$\frac{v_{ij}^{n+1} - v_{ij}^n}{\tau} = \Lambda_x v_{ij}^{n+1} + \Lambda_y v_{ij}^{n+1} + f_{ij}^{n+1}, \quad (n\tau, ih_1, jh_2) \in Q_{h,\tau},$$

$$v_{ij}^0 = \phi_{ij}, \qquad\qquad (ih_1, jh_2) \in \Omega_h, \qquad\qquad\qquad (5.32)$$

$$v_{ij}^n = \nu_{ij}^n, \qquad\qquad (n\tau, ih_1, jh_2) \in \Gamma_\tau.$$

This scheme has the local truncation error $\psi_{ij}^n = O(\tau + h_1^2 + h_2^2)$. Let us note that to find the solution v by the implicit scheme, we must solve a system of $(N_1 - 1)(N_2 - 1)$ linear equations for $n = 0, 1, ..., \frac{T}{\tau}$.

5.3.3   Crank-Nicolson Finite Difference Scheme.

Setting in (5.28) $\sigma_1 = \sigma_2 = 0.5$ and $\Theta_{ij}^n = f_{ij}^{n+0.5}$, we obtain the *Crank-Nicolson finite difference scheme*

$$\frac{v_{ij}^{n+1} - v_{ij}^n}{\tau} = \frac{1}{2}[\Lambda_x v_{ij}^{n+1} + v_{ij}^n] + \Lambda_y v_{ij}^{n+1} + v_{ij}^n] + f_{ij}^{n+0.5}, \quad (n\tau, ih_1, jh_2) \in \Gamma_{tau} \times \Omega_h,$$

$$v_{ij}^0 = \phi_{ij}, \qquad\qquad (ih_1, jh_2) \in \Omega_h,$$

$$v_{ij}^n = \nu_{ij}^n, \qquad\qquad (n\tau, ih_1, jh_2) \in \Gamma_\tau.$$

$$(5.33)$$

The Crank-Nicolson scheme has the local truncation error $\psi_{ij}^n = O(\tau^2 + h_1^2 + h_2^2)$ and it is also unconditionally convergent.

5.3.4   Estimate of the Local Truncation Error

We shall express the local truncation error in terms of the mesh steps τ, h_1 and h_2.

Namely, using Taylor's formula at the point $((n + 0.5)\tau, ih_1, jh_2)$, we find

$$u_{ij}^n = u_{ij}^{n+0.5} - \frac{\tau}{2}\frac{\partial u_{ij}^{n+0.5}}{\partial t} + \frac{\tau^2}{8}\frac{\partial^2 u_{ij}^{n+0.5}}{\partial t^2} - \frac{\tau^3}{48}\frac{\partial^3 u(\zeta_n, ih_1, jh_2)}{\partial t^3}, \qquad \zeta_n \in (t_n, t_{n+1}),$$

$$u_{ij}^{n+1} = u_{ij}^{n+0.5} + \frac{\tau}{2}\frac{\partial u_{ij}^{n+0.5}}{\partial t} + \frac{\tau^2}{8}\frac{\partial^2 u_{ij}^{n+0.5}}{\partial t^2} + \frac{\tau^3}{48}\frac{\partial^3 u(\zeta_n, ih_1, jh_2)}{\partial t^3}, \qquad \zeta_n \in (t_n, t_{n+1}),$$

$$\frac{u_{ij}^{n+1} - u_{ij}^n}{\tau} = \frac{\partial u_{ij}^{n+0.5}}{\partial t} + O(\tau^2),$$

$$\sigma_1 u_{ij}^{n+1} + (1 - \sigma_1)u_{ij}^n = u_{ij}^{n+0.5} + (2\sigma_1 - 1)\frac{\tau}{2}\frac{\partial u_{ij}^{n+0.5}}{\partial t} + \frac{\tau^2}{8}\frac{\partial^2 u_{ij}^{n+0.5}}{\partial t^2} + O(\tau^3),$$

$$\sigma_2 u_{ij}^{n+1} + (1 - \sigma_2)u_{ij}^n = u_{ij}^{n+0.5} + (2\sigma_2 - 1)\frac{\tau}{2}\frac{\partial u_{ij}^{n+0.5}}{\partial t} + \frac{\tau^2}{8}\frac{\partial^2 u_{ij}^{n+0.5}}{\partial t^2} + O(\tau^3),$$

and

$$\Lambda_x[\sigma_1 u_{ij}^{n+1} + (1 - \sigma_1)u_{ij}^n] = \frac{\partial^2 u_{ij}^{n+0.5}}{\partial x^2} + + \frac{h_1^2}{12}\frac{\partial^4 u_{ij}^{n+0.5}}{\partial x^4}$$

$$+ (2\sigma_1 - 1)\frac{\tau}{2}\Lambda_x \frac{\partial u_{ij}^{n+0.5}}{\partial t} + O(\tau^2 + h^4),$$

$$\Lambda_y[\sigma_2 u_{ij}^{n+1} + (1 - \sigma_2)u_{ij}^n] = \frac{\partial^2 u_{ij}^{n+0.5}}{\partial y^2} + \frac{h_1^2}{12}\frac{\partial^4 u_{ij}^{n+0.5}}{\partial y^4}$$

$$+ (2\sigma_2 - 1)\frac{\tau}{2}\Lambda_y \frac{\partial u_{ij}^{n+0.5}}{\partial t} + O(\tau^2 + h^4).$$

Hence

$$\psi_{ij}^n = [\frac{\partial u_{ij}^{n+0.5}}{\partial t} - \frac{\partial^2 u_{ij}^{n+0.5}}{\partial x^2} - \frac{\partial^2 u_{ij}^{n+0.5}}{\partial y^2} - \Theta_{ij}^n] - $$

$$(2\sigma_1 - 1)\frac{\tau}{2}\Lambda_x \frac{\partial u_{ij}^{n+0.5}}{\partial t} - (2\sigma_2 - 1)\frac{\tau}{2}\Lambda_y \frac{\partial u_{ij}^{n+0.5}}{\partial t} + O(\tau^2 + h^2).$$

(5.34)

Finally, the local truncation error

$$\psi_{ij}^n = \begin{cases} O(\tau^2 + h^2) & if \quad \sigma_1 = \sigma_2 = 0.5, \quad \Theta_i^n = f_i^{n+0.5}, \quad u \in C_4^{(3)}, \\ O(\tau + h^2) & if \quad \sigma_1 \neq 0.5, \quad \sigma_2 \neq 0.5, \quad \Theta_i^n = f_i^n, \quad u \in C_4^{(2)}. \end{cases}$$

(5.35)

5.3.5   Stability in the Hilbert Space $H_h^0(\Omega_h)$.

We shall investigate stability of the finite difference scheme with weight (cf. (5.28)) in the following norm:

$$||v|| = \sqrt{h_1 h_2 \sum_{i=1}^{N_1-1} \sum_{j=1}^{N_2-1} (v_{ij})^2}, \qquad v \in H_h^0(\Omega_h).$$

Let us write the scheme (5.28) in the following form:

$$(I - \tau\sigma_1\Lambda_x - \tau\sigma_2\Lambda_y)v_{ij}^{n+1} = (I + \tau(1 - \sigma_1)\Lambda_x + \tau(1 - \sigma_2)\Lambda_y)v_{ij}^n + \tau\Theta_{ij}^n, \quad (5.36)$$

where the finite difference operator

$$(I - \tau\sigma_1\Lambda_x - \tau\sigma_2\Lambda_y)$$

is positive definite in the Hilbert space $H_h^0(\Omega_h)$, Therefore, this operator is symmetric and has all eigenvalues not less than one, that is

$$\lambda_{rs} = 1 + \tau\sigma_1\lambda_r + \tau\sigma_2\lambda_s \geq 1,$$

where

$$\lambda_r = \frac{4}{h_1^2}\sin^2\frac{r\pi h_1}{2l_1}, \quad \lambda_s = \frac{4}{h_2^2}\sin^2\frac{s\pi h_2}{2l_2}, \quad r = 1,2,...,N_1-1, \;\; s = 1,2,...,N_2-1.$$

Therefore, there exists the inverse operator

$$(I - \tau\sigma_1\Lambda_x - \tau\sigma_2\Lambda_y)^{-1}.$$

and its norm

$$||(I - \tau\sigma_1\Lambda_x - \tau\sigma_2\Lambda_y)^{-1}|| \le 1.$$

Let us write scheme (5.36) in the equivalent form

$$\begin{aligned}
v_{ij}^{n+1} =\;& (I - \tau\sigma_1\Lambda_x - \tau\sigma_2\Lambda_y)^{-1}(I + \tau(1-\sigma_1)\Lambda_x + \tau(1-\sigma_2)\Lambda_y)v_{ij}^n \\
& +\tau(I - \tau\sigma_1\Lambda_x - \tau\sigma_2\Lambda_y)^{-1}\Theta_{ij}^n.
\end{aligned} \tag{5.37}$$

Multiplying both sides of scheme (5.37) by $v^{n+1} \in H_h^0(\Omega_h)$ and applying the Cauchy inequality, we obtain the following inequality:

$$\begin{aligned}
||v^{n+1}||^2 \le\;& ||(I - \tau\sigma_1\Lambda_x - \tau\sigma_2\Lambda_y)^{-1}(I + \tau(1-\sigma_1)\Lambda_x + \tau(1-\sigma_2)\Lambda_y)|| \; ||v^n|| \; ||v^{n+1}|| \\
& +\tau||(I - \tau\sigma_1\Lambda_x - \tau\sigma_2\Lambda_y)^{-1}|| \; ||\Theta^n|| \; ||v^{n+1}||.
\end{aligned}$$

Hence

$$\begin{aligned}
||v^{n+1}|| \le\;& ||(I - \tau\sigma_1\Lambda_x - \tau\sigma_2\Lambda_y)^{-1}(I + \tau(1-\sigma_1)\Lambda_x + \tau(1-\sigma_2)\Lambda_y)|| \; ||v^n|| \\
& +\tau||(I - \tau\sigma_1\Lambda_x - \tau\sigma_2\Lambda_y)^{-1}|| \; ||\Theta^n||.
\end{aligned} \tag{5.38}$$

We can get an estimate of $||v^{n+1}||$ solving inequality (5.38) under the stability condition

$$\begin{aligned}
& ||(I - \tau\sigma_1\Lambda_x - \tau\sigma_2\Lambda_y)^{-1}(I + \tau(1-\sigma_1)\Lambda_x + \tau(1-\sigma_2)\Lambda_y)|| \\
& = \max_{s,r} |\frac{(1 - \tau(1-\sigma_1)\lambda_r - \tau(1-\sigma_2)\lambda_s}{1 + \tau\sigma_1\lambda_r + \tau\sigma_2\lambda_s}| \le 1
\end{aligned} \tag{5.39}$$

The inequality (5.39) holds, if

$$-1 \le \frac{(1 - \tau(1-\sigma_1)\lambda_r - \tau(1-\sigma_2)\lambda_s}{1 + \tau\sigma_1\lambda_r + \tau\sigma_2\lambda_s} \le 1.$$

for $r = 1,2,...,N_1-1, \;\; s = 1,2,...,N_2-1$.
Hence

$$(1-2\sigma_1)\lambda_r+(1-2\sigma_2)\lambda_s \le \frac{2}{\tau}, \quad r = 1,2,...,N_1-1, \;\; s = 1,2,...,N_2-1. \tag{5.40}$$

Solving inequality (5.40), we obtain the following stability conditions:

- $\sigma_1 \geq \frac{1}{2}$ and $\sigma_2 \geq \frac{1}{2}$,

- $\sigma_1 \geq \frac{1}{2} - \frac{h_1^2}{4\tau}$ and $\sigma_2 \geq \frac{1}{2}$,

- $\sigma_1 \geq \frac{1}{2}$ and $\sigma_2 \geq \frac{1}{2} - \frac{h_2^2}{4\tau}$,

- $\sigma_1 < \frac{1}{2}$, $\sigma_2 < \frac{1}{2}$ and $\sigma_1 + \sigma_2 \geq \frac{1}{2} - \frac{h^2}{4\tau}$, $h = \max\{h_1, h_2\}$.

If one of the above conditions holds, then by inequality (5.38), we get the inequality

$$||v^n|| \leq ||v^0|| + \tau(||\Theta^0|| + ||\Theta^1|| + \cdots + ||\Theta^{\frac{T}{\tau}}||),$$

for $n = 0, 1, ..., \frac{T}{\tau}$.

Hence, we obtain the following estimate:

$$||v^n|| \leq ||\phi|| + T \max_{0 \leq n \leq \frac{T}{\tau}} |\Theta^n|, \tag{5.41}$$

for $n = 0, 1, ... \frac{T}{\tau}$.

Let us note that

- the explicit finite difference scheme is stable under the condition $\frac{\tau}{h^2} \leq \frac{1}{2}$.

- the pure explicit finite difference scheme is unconditionally stable,

- Crank-Nicolson scheme is also unconditionally stable.

5.3.6   Convergence.

In order to prove convergence of the finite difference scheme (5.28), we shall show that the global error of the method $z_{ij}^n = v_{ij}^n - u_{ij}^n$ tends to zero when $\tau, h_1, h_2 \rightarrow 0$. Indeed, error z satisfies the following equation

$$\frac{z_{ij}^{n+1} - z_{ij}^n}{\tau} = \Lambda_x[\sigma_1 z_{ij}^{n+1} + (1 - \sigma_1)z_{ij}^n] + \Lambda_y[\sigma_2 z_{ij}^{n+1} + (1 - \sigma_2)z_{ij}^n] + \psi_{ij}^n, \tag{5.42}$$

for $(n\tau, ih_1, jh_2) \in Q_{h,\tau}$, with the homogeneous initial-boundary conditions. Under the stability conditions (1), (2) or (3), error z satisfies inequality (5.41), i.e.,

$$||z^n|| \leq T \max_{0 \leq n \leq \frac{T}{\tau}} ||\psi^n||.$$

Since

$$\max_{0 \leq n \leq \frac{T}{\tau}} ||\psi^n|| \leq K(\tau + h^2),$$

when $\sigma_1 \neq \frac{1}{2}$ and $\sigma_2 \neq \frac{1}{2}$, and

$$\max_{0 \leq n \leq \frac{T}{\tau}} ||\psi^n|| \leq K(\tau^2 + h^2),$$

when $\sigma_1 = \frac{1}{2}$ and $\sigma_2 = \frac{1}{2}$,
therefore the finite difference scheme with weight converges as fast as the local truncation error ψ_{ij}^n tends to zero when $\tau, h_1, h_2 \to 0$.
Thus, the explicit finite difference scheme (5.30) is conditionally convergent, while the pure implicit and Crank-Nicolson's schemes are unconditionally convergent.

5.4   The Method of Lines for the Heat Equation

Let us consider the heat equation (5.1) with the initial boundary conditions (5.2) and (5.3). Substituting into equation (5.1) $a = 1$ and

$$\frac{\partial^2 u(t, x_i)}{\partial x^2} = \Lambda u_i(t) - \frac{h^2}{12} \frac{\partial^4 u(t, \xi_i)}{\partial x^4}, \qquad \xi_i \in (x_{i-1}, x_{i+1}),$$

we obtain the following semi discrete scheme

$$\frac{du_i(t)}{dt} = \Lambda u_i(t) + f_i(t) + E(h, t), \quad i = 1, 2, ..., N-1, \quad t \geq 0, \qquad (5.43)$$

where

$$\Lambda u_i(t) = \frac{u_{i-1}(t) - 2u_i(t) + u_{i+1}(t)}{h^2}, \qquad u_i(t) = u(t, x_i), \ x_i = ih, \ Nh = l,$$

and the local truncation error

$$E(h, t) = -\frac{h^2}{12} \frac{\partial^4 u(t, \xi_i)}{\partial x^4}.$$

Cancelling the truncation error, we get the system of ordinary differential equations

$$\frac{dv_i(t)}{dt} = \Lambda v_i(t) + f_i(t), \qquad i = 1, 2, ..., N-1, \qquad (5.44)$$

with the initial condition

$$v_i(0) = \phi(ih), \quad i = 1, 2, ..., N-1.$$

Clearly, by boundary conditions, we have

$$v_0(t) = \nu_0(t), \quad v_N(t) = \nu_N(t), \quad t \geq 0.$$

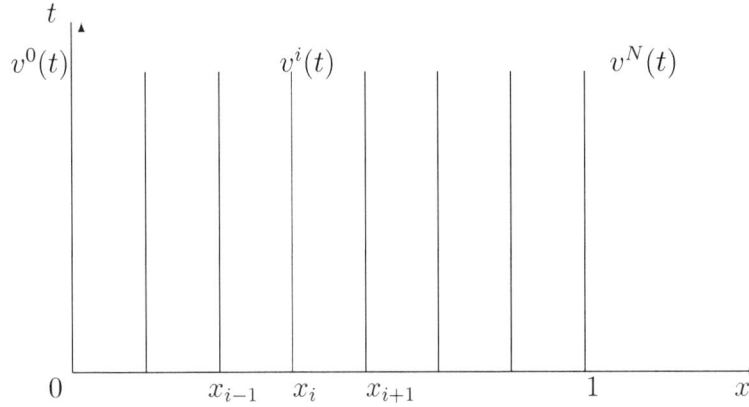

Fig. 5.2.   The method of line -network.

Convergence. We shall investigate convergence of the semidiscrete scheme (5.44) in the Hilbert space H_h^0. It is easy to check that the error of the method

$$z_i(t) = u_i(t) - v_i(t), \quad t \geq 0,$$

satisfies the following system of differential equations:

$$\frac{dz_i}{dt} = \Lambda z_i + E_i(h, t), \quad i = 1, 2, ..., N - 1, \ t \geq 0, \tag{5.45}$$

with the homogeneous initial boundary value conditions

$$z_i(0) = 0, \quad i = 1, 2, ..., N - 1,$$

$$z_0(t) = 0, \quad z_N(t) = 0, \quad t \geq 0.$$

Let us write the equations (5.45) in the vector form

$$\frac{dZ}{dt} = \Lambda Z + E(h, t), \quad i = 1, 2, ..., N - 1, \ t \geq 0,$$

where $Z(t) = [z_1(t), z_2(t), ..., z_{N-1}(t)] \in H_h^0$.
Multiplying both hand sides of equations (5.45) by Z in the sense of the inner product of the Hilbert space H_h^0, we obtain the following equation:

$$\frac{1}{2}\frac{d\|Z\|^2(t)}{dt} = (\Lambda Z(t), Z(t)) + (E(h, t), Z(t)), \quad t \geq 0, \tag{5.46}$$

with the homogeneous initial condition

$$Z(0) = 0.$$

By lemma (3.1), we have

$$(\Lambda Z(t), Z(t)) \leq -\frac{8}{l^2}(Z(t), Z(t)) + (E(h, t), Z(t)), \quad t \geq 0.$$

Hence, by (5.46) and the Cauchy inequality

$$\frac{1}{2}\frac{d||Z(t||^2(t)}{dt} \leq -\alpha ||Z||^2(t) + 2||E(h, t)|| \, ||Z||(t)), \quad \alpha = \frac{8}{l^2}, \quad t \geq 0. \quad (5.47)$$

Applying the ϵ inequality

$$|ab| \leq \epsilon a^2 + \frac{1}{4\epsilon}b^2, \quad \epsilon > 0,$$

to (5.47), we get

$$\frac{d||Z(t||^2(t)}{dt} \leq 2(\epsilon - \alpha)||Z||^2(t) + \frac{1}{2\epsilon}||E(h, t)||^2, \quad t \geq 0.$$

Choosing $\epsilon > 0$ so small to be $2(\epsilon - \alpha) = -\gamma < 0$, we find

$$||Z||^2 \leq \frac{1}{2\epsilon}\int_0^t e^{-\gamma(t-\tau)}||E(h, \tau)||^2 d\tau, \quad t > 0. \quad (5.48)$$

Because

$$||E(h, t)|| \leq \frac{h^2}{12}M^{(4)}, \quad M^{(4)} = \sup_{0 \leq x \leq l}|\frac{\partial^4 u(t, x)}{\partial x^4}|,$$

therefore, by (5.48), the error $Z(t)$ satisfies the inequality:

$$||Z||(t) \leq \frac{M^{(4)}}{12\sqrt{2\epsilon\gamma}}h^2, \quad t > 0. \quad (5.49)$$

Thus, the error $Z(t) \to 0$ as fast as $h^2 \to 0$ for every $t > 0$, so that, the method is convergent in the norm of the Hilbert space H_h^0.
In the next section, we shall prove uniform convergence of the method of lines in the maximum norm for the general model of parabolic equations.

5.4.1   Solution of homogeneous equations

Now, let us solve the system of ordinary differential equations (5.44), by the Fourier series method in the case when $\nu_0(t) = 0$, $\nu_N(t) = 0$ and $f(t, x) = 0$. Substituting into (5.44)

$$v_i(t) = T(t)V_i, \quad i = 1, 2, ..., N - 1, \ t \geq 0,$$

we obtain the following equations for $T(t)$ and V_i:

$$\frac{1}{T}\frac{dT}{dt} = -\lambda, \quad t \geq 0, \quad (5.50)$$

and

$$-\Lambda V_i = \lambda V_i, \quad i = 1, 2, ..., N-1, \quad V_0 = 0, \quad V_N = 0. \tag{5.51}$$

As we know, the solutions of the difference equation (5.51) are eigenvalues and the eigenfunctions of the operator $-\Lambda$, so that

$$\lambda_k = \frac{4}{h^2} \sin^2 \frac{k\pi h}{2l},$$

and

$$v_i^{(k)} = \sqrt{\frac{2}{l}} \sin \frac{k\pi ih}{l}, \quad i = 0, 1, ..., N, \quad k = 1, 2, ..., N-1.$$

Integrating the equation (5.50), we find

$$T_k(t) = e^{-\lambda_k t}, \quad k = 1, 2, ..., N-1, \quad t \geq 0. \tag{5.52}$$

Hence, the general solution of the equation (5.44) is:

$$v_i(t) = \sum_{k=1}^{N-1} C_k T_k(t) v_i^{(k)}, \tag{5.53}$$

where $C_k, \ k = 1, 2, ..., N-1$, are coefficients of the Fourier's series for the function $\phi(ih)$ given in the initial conditions, that is

$$v_i(0) = \phi(ih) = \sqrt{\frac{2}{l}} \sum_{k=1}^{N-1} C_k \sin \frac{k\pi ih}{l},$$

and

$$C_k = h\sqrt{\frac{2}{l}} \sum_{i=1}^{N-1} \phi(ih) \sin \frac{k\pi ih}{l}, \quad k = 1, 2, ..., N-1.$$

Example. Let us solve the heat equation

$$\frac{\partial u}{\partial t} = \frac{\partial^2 u}{\partial x^2}, \quad 0 \leq x \leq 1, \quad t > 0,$$

with the initial value condition

$$u(0, x) = x(1 - x), \quad 0 \leq x \leq 1,$$

and with the homogeneous boundary value conditions

$$u(t, 0) = 0, \quad u(t, 1) = 0, \quad t \geq 0.$$

Applying the Fourier series method, we can find the theoretical solution

$$u(t, x) = \sqrt{2} \sum_{k=1}^{\infty} c_k e^{-\sigma_k t} \sin k\pi x,$$

where $\sigma_k = \pi^2 k^2$, and the Fourier's coefficients

$$c_k = \sqrt{2} \int_0^1 x(1-x) \sin k\pi x \, dx = \frac{4\sqrt{2}}{\pi^3 k^3}[1 - (-1)^k]. \qquad (5.54)$$

By formula (5.53), the approximate solution

$$v_i(t) = \sqrt{2} \sum_{k=1}^{N-1} C_k e^{-\lambda_k t} \sin k\pi ih, \quad i = 1, 2, ..., N-1, \quad t > 0,$$

where the Fourier's coefficients

$$C_k = h\sqrt{2} \sum_{i=1}^{N-1} ih(1 - ih) \sin k\pi ih, \quad k = 1, 2, ..., N-1.$$

These coefficients are calculated by numerical integration of the integrals that appear in formula (5.54), so that

$$C_k \approx c_k = \frac{4\sqrt{2}}{\pi^3 k^3}[1 - (-1)^k], \quad k = 1, 2, ..., N-1.$$

Since $\lambda_k \approx \sigma_k$, therefore $v_i(t) \approx u(t, ih)$, $i = 1, 2, ..., N-1$, $t \geq 0$.
Solution of non-homogeneous equations. Let us consider non-homogeneous heat equation in the case when the function $f(t, x)$ has the Fourier sine series representation

$$f(t, x) = \sqrt{\frac{2}{l}} \sum_{k=1}^{\infty} f_k(t) \sin \frac{k\pi x}{l}, \quad 0 \leq x \leq l, \quad t \geq 0,$$

with the Fourier's coefficients

$$f_k(t) = \sqrt{\frac{2}{l}} \int_0^l f(t, x) \sin \frac{k\pi x}{l} dx, \quad k = 1, 2, ...$$

Then, the theoretical solution can be found in the following form

$$u(t, x) = \sqrt{\frac{2}{l}} \sum_{k=1}^{\infty} T_k(t) \sin \frac{k\pi x}{l},$$

when $u(t, 0) = 0$ and $u(t, l) = 0$, $t \geq 0$.
Indeed, substituting the sine series into (5.1), we obtain the ordinary differential equation

$$\sum_{k=1}^{\infty} [T_k''(t) + \sigma_k T_k(t) - f_k(t)] \sin \frac{k\pi x}{l} = 0, \quad \sigma_k = \frac{k^2 \pi^2}{l^2},$$

with the initial condition

$$\sqrt{\frac{2}{l}} \sum_{k=1}^{\infty} T_k(0) \sin \frac{k\pi x}{l} = \phi(x), \quad 0 \le x \le l.$$

This equation generates the decoupled system of equations

$$T_k^{''}(t) + \sigma_k T_k(t) = f_k(t), \quad k = 1, 2, ..., \quad t \ge 0,$$

with the initial conditions

$$T_k(0) = \sqrt{\frac{2}{l}} \int_0^l \phi(x) \sin \frac{k\pi x}{l} dx, \quad k = 1, 2, ...$$

Using the integrating factor, we find the following solution

$$T_k(t) = T_k(0) e^{-\sigma_k t} + \int_0^t e^{-\sigma_k(t-\tau)} f_k(\tau) d\tau.$$

Thus, the theoretical solution of the non-homogeneous heat equation is

$$u(t, x) = \sqrt{\frac{2}{l}} \sum_{k=1}^{\infty} [T_k(0) e^{-\sigma_k t} \sin \frac{k\pi x}{l} + \sin \frac{k\pi x}{l} \int_0^t e^{-\sigma_k(t-\tau)} f_k(\tau) d\tau].$$

Now, let us substitute

$$v_i(t) = \sqrt{\frac{2}{l}} \sum_{k=1}^{\infty} T_k(t) \sin \frac{k\pi i h}{l},$$

into equations (5.44). Then, following the Fourier series method, we obtain the decoupled system of equations

$$T_k^{''}(t) + \lambda_k T_k(t) = f_k(t), \quad k = 1, 2, ..., \quad t \ge 0.$$

with $\lambda_k = \dfrac{4}{h^2} \sin^2 \dfrac{k\pi h}{2l}$, and with the initial conditions

$$T_k(0) = \sqrt{\frac{2}{l}} \int_0^l \phi(x) \sin \frac{k\pi i h}{l} dx, \quad k = 1, 2, ...,$$

Also, by using the integrating factor, we find

$$T_k(t) = T_k(0) e^{-\lambda_k t} + \int_0^t e^{-\lambda_k(t-\tau)} f_k(\tau) d\tau.$$

Hence, the approximate solution of the non-homogeneous heat equation is

$$v_i(t) = \sqrt{\frac{2}{l}} \sum_{k=1}^{\infty} [T_k(0) e^{-\lambda_k t} \sin \frac{k\pi i h}{l} + \sin \frac{k\pi i h}{l} \int_0^t e^{-\lambda_k(t-\tau)} f_k(\tau) d\tau], \quad t \ge 0,$$

for $i = 1, 2, ..., N - 1$.

Example. Let us solve the initial-boundary problem

$$\frac{\partial u(t, x)}{\partial t} = \frac{\partial^2 u(t, x)}{\partial x^2} + \sin 3\pi x, \quad 0 \le x \le 1, \; t \ge 0,$$

$$u(0, x) = x(1 - x), \quad 0 \le x \le 1,$$

$$u(t, 0) = 0, \quad u(t, l) = 0, \quad t \ge 0.$$

For $f(t, x) = \sin 3\pi x$ and $\phi(x) = x(1 - x)$, $0 \le x \le 1$, $t \ge 0$, we calculate

$$f_k(t) = 2 \int_0^1 \sin k\pi x \sin 3\pi x \, dx = \begin{cases} 1, & k = 3, \\ 0, & k \ne 3, \end{cases}$$

$$T_k(0) = \sqrt{2} \int_0^1 x(1 - x) \sin k\pi x \, dx = \frac{2\sqrt{2}}{\pi^3} \frac{[1 - (-1)^k]}{k^3},$$

$$T_k(t) = \frac{4}{\pi^3} \sum_{k=1}^{\infty} \frac{[1 - (-1)^k]}{k^3} e^{\sigma_k t} + \frac{1}{\sigma_3}[1 - e^{-\sigma_3 t}].$$

Hence, the theoretical solution

$$u(t, x) = \frac{4}{\pi^3} \sum_{k=1}^{\infty} \frac{[1 - (-1)^k]}{k^3} e^{-\sigma_k t} \sin k\pi x + \frac{1}{\sigma_3}[1 - e^{-\sigma_3 t}] \sin 3\pi x.$$

In a similar way, we find the approximate solution

$$v_i(t) = \sqrt{2} \sum_{k=1}^{N-1} C_k e^{-\lambda_k t} \sin k\pi ih + \frac{1}{\lambda_3}[1 - e^{-\lambda_3 t}] \sin 3\pi ih, \quad i = 1, 2, ..., N - 1,$$

where the Fourier's coefficients

$$C_k = h\sqrt{2} \sum_{i=1}^{N-1} ih(1 - ih) \sin k\pi ih, \quad k = 1, 2, ..., N - 1.$$

Let us note that $\lambda_k \approx \sigma_k$ and $C_k \approx T_k(0)$, therefore $v_i(t) \approx u(t, ih)$, $i = 1, 2, ..., N = 1$.

5.5 Exercises

Question 5.1 *Consider the following initial-boundary value problem:*

$$\frac{\partial u}{\partial t} = 4\frac{\partial^2 u}{\partial x^2} + \sin \pi t x, \quad 0 \le x \le 1, \; t \ge 0,$$

$$u(0, x) = 0, \quad 0 \le x \le 1, \quad u(t, 0) = 0, \quad u(t, 1) = 0, \quad t \ge 0. \tag{5.55}$$

1. (a) Write down a finite difference scheme with a weight to approximate the initial-boundary problem (5.55) and determine the local truncation error.

    (b) Give the explicit finite difference scheme and find a condition under which the explicit scheme is convergent.

    (c) Give the pure implicit finite difference scheme and estimate the error of the method.

Question 5.2 Let $u(t, x)$ be solution of the heat equation

$$\frac{\partial u}{\partial t} = 8\frac{\partial^2 u}{\partial x^2} + \cos \pi t x, \quad 0 \le x \le 1, \quad t \ge 0,$$

which satisfies the initial-boundary conditions

$$u(0, x) = 0, \quad 0 \le x \le 1, \quad u(t, 0) = 0, \quad u(t, 1) = 0, \quad t \ge 0.$$

1. (a) Write down Crank-Nicolson finite difference scheme to determine approximate solution to $u(t, x)$.

    (b) Show that Crank-Nicolson finite difference scheme is unconditionally convergent.

Question 5.3 Consider the following initial boundary problem:

$$\frac{\partial u(t, x)}{\partial t} = \frac{\partial^2 u(t, x)}{\partial x^2} + x(1 - x), \quad 0 \le x \le 2, \quad t \ge 0,$$

$$u(0, x) = \sin \frac{\pi x}{2}, \quad 0 \le x \le 2,$$

$$u(t, 0) = 0, \quad u(t, 2) = 0, \quad t \ge 0.$$

1. (a) Find the theoretical solution $u(t, x)$ using the Fourier series method.

    (b) Solve the initial boundary problem by the method of lines.

    (c) Give an estimate of the error of the method of lines.

Chapter 6

Hyperbolic Equations

Tadeusz Styš
University of Warsaw

Abstract: In chapter 6, the finite difference scheme with weight $0 \leq \sigma \leq 1$ has been built for the wave equation with initial boundary value conditions. It is proved that the scheme is convergent and the global error estimate is given. The scheme with weight σ is solved by the method of separation of variables. The *Mathematica* module *waveEqn* is designed and applied to initial boundary value problems for the wave equation. In the last section, the wave equation is solved by the method of lines. The chapter ends with a set of questions.

Chapter 6

Hyperbolic Equations

6.1   Introduction

Hyperbolic equations are used in modeling of various problems arising in wave mechanics, gas dynamics, vibration, neutron diffusion, radiation, and other areas. The simplest model of those is the wave equation

$$\frac{\partial^2 u(t,x)}{\partial t^2} = \frac{\partial^2 u(t,x)}{\partial x^2} + f(t,x), \quad t \geq 0, \quad 0 \leq x \leq l, \tag{6.1}$$

with the initial conditions

$$u(0,x) = \phi_0(x), \qquad \frac{\partial u(0,x)}{\partial t} = \phi_1(x), \quad 0 \leq x \leq l, \tag{6.2}$$

and with the boundary conditions

$$u(t,0) = \nu_0(t), \qquad u(t,l) = \nu_N(t), \quad t \geq 0, \tag{6.3}$$

where $f(t,x)$, $\phi_0(x)$, $\phi_1(x)$, $\nu_0(t)$ and $\nu_1(t)$ are given continuous functions for $t \geq 0$ and $0 \leq x \leq l$.

We shall solve the initial-boundary value problem by the finite difference scheme with weight using network $Q_{h,\tau}$.

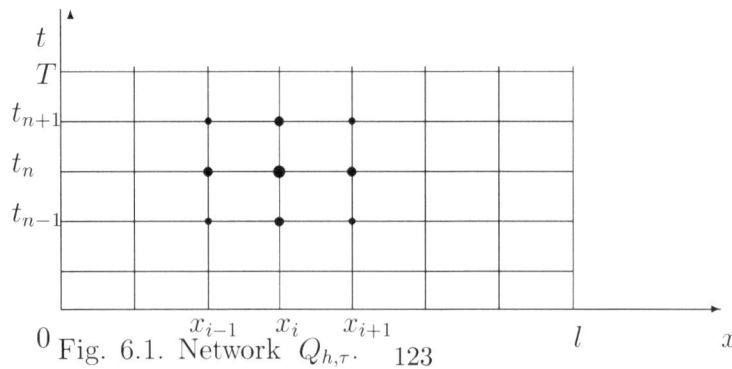

Fig. 6.1. Network $Q_{h,\tau}$.   123

The finite difference discretization of hyperbolic equations leads two tri-level schemes. We shall study such finite difference schemes in the following sections.

6.2   Finite Difference Scheme with Weight

Clearly, we can write the wave equation (6.1) in the following form:

$$\frac{\partial^2 u(t,x)}{\partial t^2} = [\sigma\frac{\partial^2 u(t,x)}{\partial x^2} + (1-2\sigma)\frac{\partial^2 u(t,x)}{\partial x^2} + \sigma\frac{\partial u(t,x)}{\partial x^2}] + f(t,x), \quad (6.4)$$

where the weight $0 \le \sigma \le 1$.

To approximate this equation, we replace the second derivatives by the corresponding finite differences

$$\Lambda_t u_i^n \equiv \frac{1}{\tau^2}[u_i^{n+1} - 2u_i^n + u_i^{n-1}] = \frac{\partial^2 u_i^n}{\partial t^2} + \frac{\tau^2}{12}\frac{\partial^4 u(\zeta_n, ih)}{\partial x^4},$$

$$\Lambda_x u_i^n \equiv \frac{1}{h^2}[u_{i-1}^n - 2u_i^n + u_{i+1}^n] = \frac{\partial^2 u_i^n}{\partial x^2} + \frac{h^2}{12}\frac{\partial^4 u(n\tau, \xi_i)}{\partial x^4},$$

$$\Lambda_x u_i^{n+1} \equiv \frac{1}{h^2}[u_{i-1}^{n+1} - 2u_i^{n+1} + u_{i+1}^{n+1}] = \frac{\partial^2 u_i^{n+1}}{\partial x^2} + \frac{h^2}{12}\frac{\partial^4 u((n+1)\tau, \xi_i)}{\partial x^4},$$

$$\Lambda_x u_i^{n-1} \equiv \frac{1}{h^2}[u_{i-1}^{n-1} - 2u_i^{n-1} + u_{i+1}^{n-1}] = \frac{\partial^2 u_i^{n-1}}{\partial x^2} + \frac{h^2}{12}\frac{\partial^4 u((n-1)\tau, \xi_i)}{\partial x^4},$$

$$(6.5)$$

for certain $\zeta_n \in (t_{n-1}, t_{n+1})$ and $\xi_i \in (x_{i-1}, x_{i+1})$, where

$$t_n = n\tau, \quad n = 0, 1, ..., \frac{T}{\tau}, \quad x_i = ih, \quad i = 0, 1, ..., N, \quad Nh = l,$$

$$\Omega_h = \{ih : i = 1, 2, ..., N-1\}, \quad Q_{h,\tau} = \Omega_h \times \{1, 2, ..., \frac{T}{\tau}\}, \quad u_i^n = u(n\tau, ih).$$

For discretization of the initial condition (6.2), we can use the following equations:

$$\frac{u(\tau, ih) - u(0, ih)}{\tau} = \frac{\partial u(0, ih)}{\partial t} + \frac{\tau}{2}\frac{\partial^2 u(0, ih)}{\partial t^2} + O(\tau^2)$$

$$= \phi_1(ih) + \frac{\tau}{2}[\frac{\partial u(0, ih)}{\partial x^2} + f(0, ih)] + O(\tau^2)$$

$$= \phi_1(ih) + \frac{\tau}{2}[\frac{d^2\phi_0(ih)}{dx^2} + f(0, ih)] + O(\tau^2).$$

Hence

$$\frac{\partial u(0, ih)}{\partial t} \approx \phi_1(ih) + \frac{\tau}{2}[\frac{d^2\phi_0(ih)}{dx^2} + f(0, ih)],$$

where the local truncation error

$$\psi_i^n(\tau) = O(\tau^2). \quad (6.6)$$

Thus, the theoretical solution $u(t,x)$ satisfies the equation

$$\frac{u(\tau, ih) - u(0, ih)}{\tau} = \phi_1(ih) + \frac{\tau}{2}\left[\frac{d^2\phi_0(ih)}{dx^2} + f(0, ih)\right] - \frac{\partial u(0, ih)}{\partial t} + \psi_i^n(\tau) + O(\tau^2).$$

From (6.4), (6.5) and (6.6), we obtain the following finite difference equation:

$$\Lambda_t u_i^n = \Lambda_x[\sigma u_i^{n+1} + (1 - 2\sigma)u_i^n + \sigma u_i^{n-1}] + \Theta_i^n + \psi_i^n(\tau, h), \qquad (6.7)$$

for $(n\tau, ih) \in Q_{h,\tau}$, where Θ_i^n depends on $f(t, x)$, and the local truncation error

$$\psi_i^n(\tau, h) = \begin{cases} \Lambda_x[\sigma u_i^{n+1} + (1 - 2\sigma)u_i^n + \sigma u_i^{n-1}] - \Lambda_t u_i^n + \Theta_i^n, & (n\tau, ih) \in Q_{h,\tau} \\ O(\tau^2), & n = 0, \quad i = 0, 1, ..., N, \\ 0, & i = 0, N, \quad n = 0, 1, ..., \frac{T}{\tau}. \end{cases}$$

$$(6.8)$$

Canceling the local truncation error in equation (6.7), we get the following finite difference scheme with weight σ (cf. [25]):

$$\begin{aligned} \Lambda_t v_i^n &= \Lambda_x[\sigma v_i^{n+1} + (1 - 2\sigma)v_i^n + \sigma v_i^{n-1}] + \Theta_i^n, & (n\tau, ih) \in Q_{h,\tau}, \\ v_i^0 &= \phi_0(ih), & i = 0, 1, ..., N, \\ \frac{v_i^1 - v_i^0}{\tau} &= \phi_1(ih) + \frac{\tau}{2}\left[\frac{d^2\phi_0(ih)}{dx^2} + f(0, ih)\right], & i = 0, 1, ..., N, \\ v_0^n &= \nu_0^n, \quad v_N^n = \nu_1^n, & n = 0, 1, ...\frac{T}{\tau}. \end{aligned}$$

$$(6.9)$$

6.2.1   Estimate of Local Truncation Error

Using the identities

$$u_i^{n+1} = u_i^n + \tau \frac{u_i^{n+1} - u_i^n}{\tau}, \qquad u_i^{n-1} = u_i^n - \tau \frac{u_i^n - u_i^{n-1}}{\tau}.$$

we determine

$$\sigma u_i^{n+1} + (1 - 2\sigma)u_i^n + \sigma u_i^{n-1} = u_i^n + \sigma\tau\left(\frac{u_i^{n+1} - u_i^n}{\tau} - \frac{u_i^n - u_i^{n-1}}{\tau}\right) = u_i^n + \sigma\tau^2\Lambda_t u_i^n.$$

By (6.8), the local error is

$$\begin{aligned} \psi_i^n(\tau, h) &= \Lambda_x u_i^n + \sigma\tau^2\Lambda_x\Lambda_t u_i^n + \Theta_i^n - \Lambda_t u_i^n \\ &= \frac{\partial u_i^n}{\partial x^2} + \frac{h^2}{12}\frac{\partial^4 u_i^n}{\partial x^4} + \sigma\tau^2\frac{\partial^4 u_i^n}{\partial x^2\partial t^2} + \Theta_i^n - \frac{\partial^2 u_i^n}{\partial t^2} + O(\tau^2 + h^4) \\ &= (\Theta_i^n - f_i^n) + (\sigma\tau^2 + \frac{h^2}{12})\frac{\partial^4 u_i^n}{\partial x^4} + \sigma\tau^2\frac{\partial^2 f_i^n}{\partial x^2} + O(\tau^2 + h^4). \end{aligned}$$

Hence, we get

$$\psi_i^n(\tau, h) = \begin{cases} O(\tau^2 + h^2) & \text{if } \sigma \neq -\dfrac{h^2}{12\tau^2}, \ \Theta_i^n = f_i^n - \sigma\tau^2\Lambda_x f_i^n, \ (n\tau, ih) \in Q_{h,\tau} \\[2mm] O(\tau^2 + h^4) & \text{if } \sigma = -\dfrac{h^2}{12\tau^2}, \ \Theta_i^n = f_i^n - \sigma\tau^2\Lambda_x f_i^n, \ (n\tau, ih) \in Q_{h,\tau}, \\[2mm] O(\tau^2), & n = 0; \quad i = 0, 1, ..., N, \\[2mm] 0, & i = 0, N; \quad n = 0, 1, ..., \dfrac{T}{\tau}. \end{cases}$$

(6.10)

6.2.2 Stability and Solution

We shall investigate stability of the finite difference scheme (6.9) with weight σ in the Hilbert space H_h^0.

Let us split the solution v, when $\nu_0(t) = 0$ and $\nu_1(t) = 0$, in two parts as follows:

$$v = V + \overline{V},$$

where V is the solution of the following homogeneous finite difference scheme ($\Theta_i^n = 0$, $\nu_0^n = 0$ and $\nu_N^n = 0$)

$$\Lambda_t v_i^n = \Lambda_x[\sigma v_i^{n+1} + (1 - 2\sigma)v_i^n + \sigma v_i^{n-1}], \quad (n\tau, ih) \in Q_{h,\tau},$$

$$v_i^0 = \phi_0(ih), \qquad\qquad\qquad\qquad i = 0, 1, ..., N,$$

$$\frac{v_i^1 - v_i^0}{\tau} = \phi_1(ih) + \frac{\tau}{2}\Big[\frac{d^2\phi_0(ih)}{dx^2} + f(0, ih)\Big] \quad i = 0, 1, ..., N,$$

$$v_0^n = 0, \quad v_N^n = 0, \qquad\qquad\qquad n = 0, 1, ...\frac{T}{\tau},$$

(6.11)

and $\overline{V}$ is the solution of the non-homogeneous finite difference scheme

$$\Lambda_t v_i^n = \Lambda_x[\sigma v_i^{n+1} + (1 - 2\sigma)v_i^n + \sigma v_i^{n-1}] + \Theta_i^n, \quad (n\tau, ih) \in Q_{h,\tau}, \qquad (6.12)$$

with the homogeneous initial-boundary conditions

$$v_i^0 = 0, \quad v_i^1 = 0 \quad i = 0, 1, ..., N,$$

$$v_0^n = 0, \quad v_N^n = 0, \quad n = 0, 1, ...\frac{T}{\tau}.$$

(6.13)

We shall find V solving equation (6.11) by the method of variables separation, so that, substituting to (6.11)

$$V_i^n = W_i T^n \neq 0.$$

we obtain

$$\frac{\Lambda_x W_i}{W_i} = \frac{\Lambda_t T^n}{\sigma T^{n+1} + (1 - 2\sigma)T^n + \sigma T^{n-1}} = -\lambda, \quad (n\tau, ih) \in Q_{h,\tau}. \qquad (6.14)$$

Hence

$$\Lambda_x W_i + \lambda W_i = 0, \quad i = 1, 2, ..., N - 1,$$

$$W_0 = 0, \quad W_N = 0.$$

(6.15)

Thus, the eigenvalues and eigenfunctions of the difference operator $-\Lambda_x$ are:

$$\lambda_k = \frac{4}{h^2} sin^2 \frac{k\pi h}{2l}, \quad W_i^k = \sqrt{\frac{2}{l}} sin \frac{k\pi ih}{l}, \quad k = 1, 2, ..., N - 1.$$

Also, from (6.14), the discrete function $T^n \neq 0$ satisfies the difference equation

$$\Lambda_t T^n + \lambda_k [\sigma T^{n+1} + (1 - 2\sigma)T^n + \sigma T^{n-1}] = 0, \quad n = 1, 2, ..., \frac{T}{\tau}. \quad (6.16)$$

Let us rewrite this equation in the following form:

$$(1 + \sigma \tau^2 \lambda_k)T^{n+1} - (2 + (2\sigma - 1)\tau^2 \lambda_k)T^n + (1 + \sigma \tau^2 \lambda_k)T^{n-1} = 0,$$

or

$$T^{n+1} - \alpha_k T^n + T^{n-1} = 0, \quad n = 1, 2, ..., \frac{T}{\tau}, \quad (6.17)$$

where

$$\alpha_k = \frac{2 + (2\sigma - 1)\tau^2 \lambda_k}{1 + \sigma \tau^2 \lambda_k}.$$

One can check that

$$-2 < \alpha_k < 2 \quad \text{for} \quad \frac{\tau^2}{h^2} \leq 1 \quad \text{and} \quad 0 \leq \sigma \leq 1. \quad (6.18)$$

Therefore, the characteristic equation

$$\xi^2 - \alpha_k \xi + 1 = 0$$

of the difference equation (6.17) has the following complex roots:

$$\xi_1 = e^{i\beta_k}, \quad \xi_2 = e^{-i\beta_k}, \quad \text{where} \quad \beta_k = Arg(\frac{\alpha_k + i\sqrt{4 - \alpha_k^2}}{2}).$$

Under condition (6.18), all solutions of the difference equation (6.17) are bounded, and its general solution is:

$$T_k^n = a_k \cos n\beta_k + b_k \sin n\beta_k, \quad n = 1, 2, ..., \frac{T}{\tau}, \quad (6.19)$$

where a_k and b_k, $k = 1, 2, ..., N - 1$, are arbitrary constants.
Now, we can present V as the following sum of the particular solutions $W_i^k T_k^n$:

$$V_i^n = \sum_{k=1}^{N-1} [a_k \cos n\beta_k + b_k \sin \beta_k] W_i^k. \quad (6.20)$$

Hence, by the initial conditions

$$V_i^0 = \phi_0(ih) = \sum_{k=1}^{N-1} \phi_0^k W_i^k \quad \text{and} \quad \frac{V_i^1 - V_i^0}{\tau} = \overline{\phi}_1(ih) = \sum_{k=1}^{N-1} \overline{\phi}_1^k W_i^k, \qquad (6.21)$$

where ϕ_0^k, $\overline{\phi}_1^k$, $k = 1, 2, ..., N-1$, are Fourier's coefficients of the functions $\phi_0(ih)$ and

$$\overline{\phi}_1(ih) = \phi_1(ih) + \frac{\tau}{2}\left[\frac{d^2\phi_0(ih)}{dx^2} + f(0, ih)\right],$$

i.e. $\phi_0^k = (\phi_0, W^k)$ and $\overline{\phi}_1^k = (\overline{\phi}_1, W^k)$, for $k = 1, 2, ..., N-1$. Comparing (6.20) and (6.21) for $n = 0$, we find

$$a_k = \phi_0^k \qquad \text{and} \qquad a_k \frac{\cos \beta_k - 1}{\tau} + b_k \frac{\sin \beta_k}{\tau} = \overline{\phi}_1^k.$$

Hence

$$a_k = \phi_0^k, \qquad b_k = \frac{1 - \cos \beta_k}{\sin \beta_k}\phi_0^k + \frac{\tau}{\sin \beta_k}\overline{\phi}_1^k.$$

Finally, the solution of the initial-boundary problem (6.11) is:

$$V_i^n = \sum_{k=1}^{N-1}\left[\frac{\cos (n - 0.5)\beta_k}{\cos 0.5\beta_k}\phi_0^k + \frac{\tau \sin n\beta_k}{\sin \beta_k}\overline{\phi}_1^k\right]W_i^k, \quad (n\tau, ih) \in Q_{h,\tau}. \qquad (6.22)$$

In order to estimate V^n in the norm of the Hilbert space H_h^0, we calculate

$$||V^n||^2 = h\sum_{i=1}^{N-1} V_i^n V_i^n = h\sum_{i=1}^{N-1}\left\{\sum_{k=1}^{N-1}\left[\frac{\cos(n - 0.5)\beta_k}{\cos 0.5\beta_k}\phi_0^k + \frac{\tau \sin n\beta_k}{\sin \beta_k}\overline{\phi}_1^k\right]W_i^k\right\}^2 =$$

$$= \sum_{k=1}^{N-1}\left[\frac{\cos(n - 0.5)\beta_k}{\cos 0.5\beta_k}\phi_0^k + \frac{\tau \sin n\beta_k}{\sin \beta_k}\overline{\phi}_1^k\right]\sum_{r=1}^{N-1}\left[\frac{\cos(n - 0.5)\beta_r}{\cos 0.5\beta_r}\phi_0^r + \frac{\tau \sin n\beta_r}{\sin \beta_r}\overline{\phi}_1^r\right]\delta_{kr} =$$

$$= \sum_{k=1}^{N-1}\left[\frac{\cos(n - 0.5)\beta_k}{\cos 0.5\beta_k}\phi_0^k + \frac{\tau \sin n\beta_k}{\sin \beta_k}\overline{\phi}_1^k\right]^2,$$

where $\delta_{kr} = h\sum_{i=1}^{N-1} W_i^k W_i^l$ is the Kronecker's delta.

Hence, by the inequality $(a + b)^2 \leq 2(a^2 + b^2)$, we get

$$||V^n||^2 \leq 2\sum_{k=1}^{N-1}\left[\frac{1}{\cos^2 0.5\beta_k}(\phi_0^k)^2 + \left(\frac{\tau}{\sin \beta_k}\right)^2(\overline{\phi}_1^k)^2\right] \qquad (6.23)$$

For $\dfrac{\tau^2}{h^2} \leq \gamma < 1$, we have

$$0 < \tau^2 \lambda_k = \frac{4\tau^2}{h^2} \sin \frac{k\pi h}{2l} < 4\gamma,$$

$$
\begin{aligned}
\cos^2 0.5\beta_k &= \frac{1}{2}(1 + \cos \beta_k) = \frac{1}{2}\left(1 + \frac{\alpha_k}{2}\right) \\
&= \frac{1}{2} + \frac{2 + (2\sigma - 1)\tau^2 \lambda_k}{4(1 + \sigma\tau^2\lambda_k)} \\
&\geq \frac{1}{2} + \frac{2 + (2\sigma - 1)4\gamma}{4(1 + 4\sigma\gamma)} \geq \frac{1}{2} + \frac{1}{4}(2 - 4\gamma) = 1 - \gamma > 0.
\end{aligned}
\tag{6.24}
$$

and

$$
\begin{aligned}
\sin \beta_k &= \frac{1}{2}\sqrt{4 - \alpha_k^2}, \qquad \alpha_k = 2 - \frac{\tau^2\lambda_k}{1 + \sigma\tau^2\lambda_k}, \\
\frac{\tau^2}{\sin^2 \beta_k} &= \frac{4\tau^2}{(2 + \alpha_k)(2 - \alpha_k)} = \frac{4(1 + \sigma\tau^2\lambda_k)^2}{\lambda_k(4 + 4\sigma\tau^2\lambda_k - \tau^2\lambda_k)}.
\end{aligned}
\tag{6.25}
$$

Hence, we obtain the following condition for the weight σ

$$4 + 4\sigma\,\tau^2\lambda_k - \tau^2\lambda_k > 0, \quad \text{or} \quad \sigma > \frac{1}{4} - \frac{1}{\tau^2\lambda_k}.$$

Because $\tau^2\lambda_k \leq \dfrac{4\tau^2}{h^2}$, therefore, the condition of stability is

$$\sigma > \frac{1}{4} - \frac{h^2}{4\tau^2}.\tag{6.26}$$

By the inequality $\dfrac{8}{l^2} < \lambda_k < \dfrac{4}{h^2}$, we have

$$\frac{\tau^2}{\sin^2 \beta_k} < \frac{l^2 + 8}{8(1 - \gamma)}.\tag{6.27}$$

From inequalities (6.23), (6.24) and (6.27), we obtain the following a priori estimate

$$||V^n||^2 \leq K \sum_{k=1}^{N-1} [(\phi_0^k)^2 + (\overline{\phi}_1^k)^2], \qquad n = 1, 2, ..., \frac{T}{\tau},\tag{6.28}$$

where constant

$$K = 2\max\left\{\frac{1}{1 - \gamma}, \frac{l^2 + 8}{8(1 - \gamma)}\right\}$$

is independent of the mesh steps τ and h.

Because $\{W^k\}$, $k = 1, 2, ... N - 1$, is a complete orthonormal set of discrete functions in the Hilbert space H_h^0, the Parserval equality holds for any $v \in H_h^0$, so that

$$||\phi_0||^2 = \sum_{k=1}^{N-1} (\phi_0^k)^2, \quad \text{and} \quad ||\overline{\phi}_1||^2 = \sum_{k=1}^{N-1} (\overline{\phi}_1^k)^2.$$

Then, by (6.28), we get the following a priori estimate:

$$||V^n|| \leq K[\, ||\phi_0|| + ||\overline{\phi}_1||\,], \quad n = 1, 2, ..., \frac{T}{\tau}. \tag{6.29}$$

where K is a generic constant independent of the mesh steps τ and h.

Now, let us estimate the solution $\overline{V}^n$ of the non-homogeneous finite difference scheme (6.12), (6.13). Obviously, from (6.12), the solution $\overline{V}^n$ satisfies the equation

$$(I - \sigma\tau^2\Lambda_x)\overline{V}_i^{n+1} = [2I - (2\sigma - 1)\tau^2\Lambda_x]\overline{V}_i^n - (I - \sigma\tau^2\Lambda_x)\overline{V}_i^{n-1} + \tau^2\Theta_i^n, \tag{6.30}$$

for $(n\tau, ih) \in Q_{h,\tau}$, where I is the identity operator.

Let us substitute into (6.30)

$$\overline{V}_i^n = \sum_{k=1}^{N-1} a_k^n W_i^k,$$

where the Fourier's coefficients

$$a_k^n = (\overline{V}^n, W^k) = h \sum_{i=1}^{N-1} \overline{V}_i^n W_i^k, \quad k = 1, 2, ...N - 1.$$

Then, we obtain

$$\sum_{k=1}^{N-1} (I - \sigma\tau^2\Lambda_x) a_k^{n+1} W_i^k = \sum_{k=1}^{N-1} [2I - (2\sigma - 1)\Lambda_x] a_k^n W_i^k$$
$$- \sum_{k=1}^{N-1} (I - \sigma\tau^2\Lambda_x) a_k^{n-1} W_i^k + \tau^2\Theta_i^n,$$

and

$$\sum_{k=1}^{N-1} (1 + \sigma\tau^2\lambda_k) a_k^{n+1} W_i^k = \sum_{k=1}^{N-1} \{[2 + (2\sigma - 1)\lambda_k] a_k^n$$
$$- (1 + \sigma\tau^2\lambda_x) a_k^{n-1}\} W_i^k + \tau^2\Theta_i^n. \tag{6.31}$$

Multiply both sides of (6.31) by W^r, $r = 1, 2, ...N - 1$, we obtain

$$\sum_{k=1}^{N-1} (1 + \sigma\tau^2\lambda_k) a_k^{n+1} (W^k, W^r) = \sum_{k=1}^{N-1} \{[2 - (2\sigma - 1)\lambda_k] a_k^n -$$
$$- \sum_{k=1}^{N-1} (1 + \sigma\tau^2\lambda_x) a_k^{n-1}\} (W^k, W^r) + \tau^2 (\Theta^n, W^r). \tag{6.32}$$

Because

$$(W^k, W^r) = \begin{cases} 1 & if \quad k = r, \\ 0 & if \quad k \neq r, \end{cases}$$

therefore, from (6.32)

$$(1 + \sigma\tau^2\lambda_r)a_r^{n+1} = [2 + (2\sigma - 1)\tau^2\lambda_r]a_r^n - (1 + \sigma\tau^2\lambda_r)a_{rk}^{n-1} + \tau^2(\Theta_i^n, W^r),$$

Thus, the Fourier's coefficients a_r^n, $r = 1, 2, ..., N - 1$ satisfy the following difference equation:

$$a_r^{n+2} - \alpha_r a_r^{n+1} + a_r^n = \tau^2 S_r(\Theta^{n+1}, W^r), \tag{6.33}$$

with the initial conditions

$$a_r^0 = a_r^1 = 0, \qquad r = 1, 2, ..., N - 1,$$

where

$$\alpha_r = \frac{2 + (2\sigma - 1)\tau^2\lambda_r}{1 + \sigma\tau^2\lambda_r}, \qquad S_r = \frac{1}{1 + \sigma\tau^2\lambda_r}.$$

Because $-2 < \alpha_r < 2$, $r = 1, 2, ..., N - 1$, for $\dfrac{\tau^2}{h^2} \leq \gamma < 1$, the characteristic equation of (6.33)

$$\xi^2 - \alpha_r\xi + 1 = 0$$

has the complex roots

$$\lambda_{r1} = e^{i\beta_r}, \qquad \lambda_{r2} = e^{-i\beta_r}, \qquad r = 1, 2, ..., N - 1,$$

where

$$\cos\beta_r = \frac{\alpha_r}{2} \quad and \quad \sin\beta_r = \frac{\sqrt{4 - \alpha_r^2}}{2}.$$

Therefore, the homogeneous difference equation

$$a_r^{n+2} - \alpha_r a_r^{n+1} + a_r^n = 0, \tag{6.34}$$

has the general solution

$$a_r^n = C_1\cos n\beta_r + C_2\sin n\beta_r, \qquad n = 0, 1, ...; \quad r = 1, 2, ..., N - 1. \tag{6.35}$$

In order to find a particular solution of non-homogeneous equation (6.33), we apply the method of variation of parameters. Namely, by the formula (1.18), the particular solution of (6.33)

$$a_{*r}^n = C_1^r(n)\cos n\beta_r + C_2^r(n)\sin n\beta_r, \tag{6.36}$$

where

$$C_1^r(n) = \sum_{k=0}^{n-1} \frac{D_1^r(k)}{D^r(k)}, \qquad C_2^r(n) = \sum_{k=0}^{n-1} \frac{D_2^r(k)}{D^r(k)}.$$

and

$$D^r(k) = \begin{vmatrix} \cos(k+1)\beta_r & \sin(k+1)\beta_r \\ \cos(k+2)\beta_r & \sin(k+2)\beta_r \end{vmatrix} = \sin\beta_r,$$

$$D_1^r(k) = \begin{vmatrix} 0 & \sin(k+1)\beta_r \\ \tau^2 S_r(\Theta^k, W^r) & \sin(k+2)\beta_r \end{vmatrix} = -\tau^2 S_r \sin(k+1)(\Theta^k, W^r),$$

$$D_2^r(k) = \begin{vmatrix} \cos(k+1)\beta_r & 0 \\ \cos(k+2)\beta_r & \tau^2 S_r(\Theta^k, W^r) \end{vmatrix} = \tau^2 S_r \cos(k+1)(\Theta^k, W^r).$$

for $k = 0, 1, ..., n-1$.

Hence, we find the coefficients

$$\begin{aligned} C_1^r(n) &= -\sum_{k=0}^{n-1} \frac{\tau^2 S_r(\Theta^k, W^r)}{\sin\beta_r} \sin(k+1)\beta_r, \\ C_2^r(n) &= \sum_{k=0}^{n-1} \frac{\tau^2 S_r(\Theta^k, W^r)}{\sin\beta_r} \cos(k+1)\beta_r. \end{aligned} \tag{6.37}$$

for $n = 1, 2, ...$; and $C_1^r(0) = C_2^r(0) = 0$, $r = 1, 2, ..., N-1$.

From (6.36) and (6.37), the particular solution

$$a_{*r}^n = \sum_{k=0}^{n-1} \frac{\tau^2 S_r(\Theta^k, W^r)}{\sin\beta_r}[\cos(k+1)\beta_r \sin n\beta_r - \sin(k+1)\beta_r \cos n\beta_r].$$

The general solution of (6.33) is the sum of the general solution (6.35) of the homogeneous equation (6.34) and the particular solution a_{*r}^n.

Thus, we have

$$a_r^n = \begin{cases} C_1 \cos n\beta_r + C_2 \sin n\beta_r + \sum_{k=0}^{n-1} \frac{\tau^2 S_r(\Theta^k, W^r)}{\sin\beta_r} \sin(n-k-1)\beta_r, \\ \text{for}\quad n = 1, 2, ...; \\ C_1, \quad \text{for}\quad n = 0. \end{cases} \tag{6.38}$$

Because

$$a_r^0 = 0, \qquad a_r^1 = 0, \qquad r = 1, 2, ..., N-1,$$

we get

$$C_1 = 0 \quad \text{and} \quad C_2 = 0.$$

Hence, the Fourier's coefficients are given by

$$a_r^n = \sum_{k=0}^{n-1} \frac{\tau^2 S_r(\Theta^k, W^r)}{\sin\beta_r} \sin(n-k-1)\beta_r. \quad n = 1, 2, ...; \tag{6.39}$$

From (6.38) and (6.39), we obtain the solution of the non-homogeneous difference equation (6.12)

$$\overline{V}_i^n = \sum_{r=1}^{N-1} \{ \sum_{k=0}^{n-1} \frac{\tau^2 S_r \overline{\Theta}_r^k}{\sin \beta_r} \sin(n-k-1)\beta_r \} W_i^r, \quad (n\tau, ih) \in Q_{h,\tau}. \quad (6.40)$$

where $\overline{\Theta}_r^k = (\Theta^k, W^r)$, $r = 1, 2, ..., N-1$, are Fourier coefficients of Θ^k, $k = 0, 1, ..., \frac{T}{\tau}$.

Let us estimate the solution $\overline{V}^n$ on each level $n = 0, 1, ..., \frac{T}{\tau}$.

Using Parserval's equality and the inequality (6.27), we compute the norm

$$
\begin{aligned}
||\overline{V}^n||^2 &= \sum_{r=1}^{N-1} [a_r^n]^2 \sum_{r=1}^{N-1} \sum_{k=0}^{n-1} \{ \frac{\tau^2 S_r \overline{\Theta}_r^k}{\sin \beta_r} \sin((n-k-1)\beta_r) \}^2 \\
&\leq \frac{l^2 + 8}{8(1-\gamma)} \sum_{r=1}^{N-1} \{ \sum_{k=0}^{n-1} \tau \overline{\Theta}_r^k \}^2 \\
&\leq \frac{(l^2 + 8)T^2}{8(1-\gamma)} \sum_{r=1}^{N-1} [\max_{0 \leq k < n} \overline{\Theta}_r^k]^2 \\
&= \frac{(l^2 + 8)T^2}{8(1-\gamma)} ||\Theta^{n-1}||_{max}^2
\end{aligned}
\quad (6.41)
$$

where

$$||\Theta^{n-1}||_{max}^2 = \sum_{r=1}^{N-1} [\max_{0 \leq k < n} \overline{\Theta}_r^k]^2.$$

Hence, we get the following estimate of the solution $\overline{V}$:

$$||\overline{V}^n|| \leq KT||\Theta^{n-1}||_{max}, \quad n = 0, 1, ..., \frac{T}{\tau}, \quad (6.42)$$

where $K = \sqrt{\frac{l^2 + 8}{8(1-\gamma)}}$.

Finally, by formulas (6.22) and (6.40), we obtain the solution v of the finite difference scheme (6.9)

$$
\begin{aligned}
v_i^n = V_i^n + \overline{V}_i^n = \sum_{k=1}^{N-1} [&\frac{\cos (n-0.5)\beta_k}{\cos 0.5\beta_k} \phi_0^k + \frac{\tau \sin n\beta_k}{\sin \beta_k} \overline{\phi}_1^k \\
&+ \sum_{r=0}^{n-1} \frac{\tau^2 S_r (\Theta^r, W^k)}{\sin \beta_k} \sin(n-k-1)\beta_k] W_i^k.
\end{aligned}
\quad (6.43)
$$

for $\nu_0(t) = \nu_1(t) = 0$, $t \geq 0$ and $(n\tau, ih) \in Q_{h,\tau}$.

From inequalities (6.29) and (6.42), we get the following estimate of the solution v:

$$||v^n|| \leq K[||\phi_0|| + ||\overline{\phi}_1||] + TK||\Theta^{n-1}||_{max}, \quad n = 0, 1, ..., \frac{T}{\tau}, \quad (6.44)$$

where K is a generic constant independent of τ and h.

This estimate means stability of the finite difference scheme (6.9) with the weight $0 \leq \sigma \leq 1$ in the norm of the Hilbert space H_h^0, provided that $\frac{\tau^2}{h^2} \leq \gamma < 1$.

6.2.3   Convergence

We shall show that the approximate solution $v_i^n \rightarrow u_i^n$ when $\tau, h \rightarrow 0$ at any fixed mesh point $(t_n^*, x_n^*) \in Q_{h,\tau}$, $(t_n^* = n\tau = constant, \ x_i^* = ih = constant.)$
Indeed, let

$$z_i^n = v_i^n - u_i^n, \qquad i = 0, 1, ..., N, \ \ n = 0, 1, ..., \frac{T}{\tau},$$

be the global error of the method. Obviously, z is the solution of the following finite difference scheme:

$$\Lambda_t z_i^n = \Lambda_x [\sigma z_i^{n+1} + (1 - 2\sigma) z_i^n + z_i^{n-1}] + \psi_i^n(\tau, h), \ \ (n\tau, ih) \in Q_{h,\tau},$$

$$z_i^0 = 0, \qquad \frac{z_i^1 - z_i^0}{\tau} = \psi_i^n(\tau, h) \hspace{2cm} i = 0, 1, ..., N, \qquad (6.45)$$

$$v_0^n = 0, \qquad v_N^n = 0, \hspace{3cm} n = 0, 1, ... \frac{T}{\tau},$$

where $\psi_i^n(\tau, h)$ is the local truncation error (cf. (6.10)).

From estimate (6.44), the error z satisfies the inequality

$$||z^n|| \leq KT ||\psi||_{max}, \quad n = 0, 1, ..., \frac{T}{\tau}. \hspace{2cm} (6.46)$$

Since

$$||\psi||_{max} \leq K \begin{cases} \tau^2 + h^2 \ \ if \ \ \sigma \neq -\dfrac{h^2}{12\tau}, & (n\tau, ih) \in Q_{h,\tau}, \\[2mm] \tau^2 + h^4 \ \ if \ \ \sigma = -\dfrac{h^2}{12\tau}, & (n\tau, ih) \in Q_{h,\tau}, \\[2mm] \tau^2 \hspace{1.2cm} n = 0, 1, ..., \dfrac{T}{\tau}, \\[2mm] 0, \end{cases}$$

we have

$$||z^n|| \leq K \begin{cases} \tau^2 + h^2 \ \ if \ \ \sigma \neq -\dfrac{h^2}{12\tau}, \\[2mm] \tau^2 + h^4 \ \ if \ \ \sigma = -\dfrac{h^2}{12\tau}, \end{cases}$$

for $n = 1, 2, ..., \frac{T}{\tau}$ and $\frac{\tau^2}{h^2} \leq \gamma < 1$.

Thus, the finite difference scheme (6.9) is convergent in the norm of the

Hilbert's space H_h^0 as fast as $O(\tau^2 + h^2) \to 0$ when $\sigma \neq -\dfrac{h^2}{12\tau}$, and as fast as $O(\tau^2 + h^4) \to 0$ when $\sigma = -\dfrac{h^2}{12\tau}$, provided that $\dfrac{\tau^2}{h^2} \leq \gamma < 1$.

6.2.4 Numerical Solution with Mathematica

In oder to compute a solution of the initial boundary value problem (6.1), (6.2), (6.3), we apply the finite difference scheme with weight (6.9) or the formula (6.22).

The following `Mathematica` module `waveEqn` solves the wave equation (6.1) by the finite difference scheme (6.9).

Program 6.1 *Mathematica module that solves discrete finite difference scheme with weight for wave equation*

```
waveEqn[f_,phi0_,phi1_,sigma_,nx_,nt_,tau_]:=Module[
          {c,d,f1,v0,v1,v2,sol},
h=1./nx; d=1+2*sigma (tau/h)^2;  c=-sigma(tau/h)^2;

f1[k_]:=Table[f[k*tau,i*h],{i,1,nx-1}];
v0=N[Table[phi0[i*h],{i,0,nx}]];
v1=v0+tau*N[Table[phi1[i*h]+tau(phi0''[i h]+f[0,i h])/2,
          {i,0,nx}]];

laplace1[v_,h_]:=Take[(RotateRight[v,1]-2 v +
                    RotateLeft[v,1])/h^2,{2,-2}];

soltri[k_,v0_,v1_]:=Module[{al,be,lh,x},
 lh=2*Take[v1,{2,-2}]-Take[v0,{2,-2}]+
     tau^2 (1-2 sigma)*laplace1[v1,h]+
     tau^2 sigma*laplace1[v0,h]+tau^2 f1[k];
 al[1]=c/d;
  al[i_]:=al[i]=c/(d-al[i-1]*c);
  be[1]=lh[[1]]/d;
  be[i_]:=be[i]=(lh[[i]]-be[i-1]*c)/(d-al[i-1]*c);
  x[nx-1]=be[nx-1];
  x[i_]:=x[i]=be[i]-al[i]*x[i+1];
  Append[Prepend[Table[x[i],{i,1,nx-1}],
             N[ bleft[k*tau]]],N[bright[k*tau]]]
 ];
 sol={v0,v1};
```

```
Do[{v2=soltri[j+1,v0,v1];sol=AppendTo[sol,v2];
    v0=v1;v1=v2},{j,1,9}];
sol =Chop[N[sol,3]]
]
```

Example 6.1 *let us solve the following initial boundary value problem by the module* waveEqn

$$\frac{\partial^2 u(t,x)}{\partial t^2} = \frac{\partial^2 u(t,x)}{\partial x^2}, \quad t \geq 0, \quad 0 \leq x \leq 1,$$

$$u(0,x) = e^{-x} \sin \pi x, \quad \frac{\partial u(0,x)}{\partial t} = e^{-x}(\pi \cos \pi x - \sin \pi x), \quad (6.47)$$

$$u(t,0) = e^{-t} \sin \pi t, \quad u(t,1) = e^{-(t+1)} \sin \pi(t+1).$$

The input data are:

```
f[t_,x_]:=0; phi0[x_]:=Exp[-x] Sin[Pi x];
phi1[x_]:=Exp[-x](Pi Cos[Pi x]-Sin[Pi x]);
bleft[t_]:=Exp[-t] Sin[Pi t];
bright[t_]:=Exp[-(t+1)]Sin[Pi (t+1)];
```

By executing the command waveEqn[f,phi0,phi1,0,5,9,0.1] with the parameters $\sigma = 0$, $nx = 5$, $nt = 9$, $\tau = 0.1$, we obtain the numerical solution tabulated below along with the theoretical solution $u(t,x) = e^{-(t+x)} \sin \pi(t+x)$.

t/x	Exact Solution u_i^n				Numerical Solution v_i^n, $\sigma = 0$			
	0.2	0.4	0.6	0.8	0.2	0.4	0.6	0.8
0.0	0.481	0.638	0.522	0.264	0.481	0.638	0.522	0.264
0.1	0.599	0.607	0.402	0.126	0.599	0.604	0.399	0.123
0.2	0.638	0.522	0.264	0.000	0.639	0.518	0.258	-0.006
0.3	0.607	0.402	0.126	-0.103	0.609	0.397	0.116	-0.111
0.4	0.522	0.264	0.000	-0.103	0.524	0.259	-0.012	-0.188
0.5	0.402	0.126	-0.103	-0.220	0.401	0.120	-0.117	-0.232
0.6	0.264	0.0.000	-0.177	-0.235	0.259	-0.009	-0.191	-0.245
0.7	0.126	-0.103	-0.220	-0.223	0.116	-0.116	-0.233	-0.231
0.8	0.0.000	-0.177	-0.235	-0.192	-0.114	-0.194	-0.245	-0.197
0.9	-0.103	-0.220	-0.223	-0.148	-0.119	-0.240	-0.233	-0.150
1.0	-0.177	-0.235	-0.192	-0.097	-0.193	-0.254	-0.202	-0.098
	Error: max $\mid v_i^n - u_i^n \mid = 0.0196$ when $h = 0.2$, $\tau = 0.1$							

Example 6.2 *Consider the wave equation*

$$\frac{\partial^2 u(t,x)}{\partial t^2} = \frac{\partial^2 u(t,x)}{\partial x^2}, \quad 0 < x < 1, \quad t > 0, \quad (6.48)$$

with the initial conditions

$$u(0, x) = sin\pi x, \qquad \frac{\partial u(0, x)}{\partial t} = 0, \qquad 0 \le x \le 1, \qquad (6.49)$$

and with the boundary conditions

$$u(t, 0) = 0, \qquad u(t, 1) = 0, \qquad t \ge 0. \qquad (6.50)$$

The theoretical solution of this initial-boundary value problem

$$u(t, x) = \sin \pi x \cos \pi t, \qquad 0 \le x \le 1, \qquad t \ge 0.$$

We shall solve equation (6.48) under the initial-boundary conditions (6.49), (6.50), by the finite difference scheme (6.9). Because $f(t, x) = 0$, $\phi_0(x) = \sin \pi x$, $\phi_1(x) = 0$, $\nu_0(t) = 0$ and $\nu_1(t) = 0$ for $0 \le x \le 1$, $t \ge 0$. Therefore, the approximate solution v satisfies the following finite difference scheme:

$$\Lambda_t v_i^n = \Lambda_x [\sigma v_i^{n+1} + (1 - 2\sigma) v_i^n + v_i^{n-1}], \qquad (n\tau, ih) \in Q_{h,\tau},$$

$$v_i^0 = \sin \pi ih, \qquad \frac{v_i^1 - v_i^0}{\tau} = -\frac{\tau \pi^2}{2} \sin \pi ih, \quad i = 0, 1, ..., N, \qquad (6.51)$$

$$v_0^n = 0, \qquad v_N^n = 0, \qquad\qquad\qquad n = 0, 1, ... \frac{T}{\tau}.$$

In this example, $v_i^n = V_i^n$, since $\overline{V}_i^n = 0$. From formula (6.22)

$$v_i^n = \sum_{k=1}^{N-1} [\frac{\cos (n - 0.5)\beta_k}{\cos 0.5\beta_k} \phi_0^k + \frac{\tau \sin n\beta_k}{\sin \beta_k} \overline{\phi}_1^k] W_i^k, \qquad (n\tau, ih) \in Q_{h,\tau}, \qquad (6.52)$$

where the Fourier coefficients

$$\phi_0^k = (\phi_0, W^k) = h \sum_{i=1}^{N-1} \sqrt{2} \sin \pi ih \sin k\pi ih = \begin{cases} 1 & if \quad k = 1, \\ 0 & if \quad k \ne 1. \end{cases}$$

and

$$\overline{\phi}_1^k = (\phi_1, W^k) = -\frac{\tau \pi^2}{2} \sqrt{2} h \sum_{i=1}^{N-1} \sin \pi ih \sin k\pi ih = \begin{cases} -\frac{\tau \pi^2}{2} & if \quad k = 1, \\ 0 & if \quad k \ne 1. \end{cases}$$

Hence, the solution

$$v_i^n = [\frac{\cos(n - 0.5)\beta_1}{\cos 0.5\beta_1} - \frac{\tau^2 \pi^2}{2} \frac{\sin n\beta_1}{\sin \beta_1}] \sin \pi ih, \qquad (n\tau, ih) \in Q_{h,\tau}. \qquad (6.53)$$

and the numerical results are given in the following table:

	Exact Solution u_i^n				Numerical Solution v_i^n, $\sigma = 1$			
t/x	0.2	0.4	0.6	0.8	0.2	0.4	0.6	0.8
0.1	0.309	0.588	0.809	0.951	0.309	0.588	0.809	0.951
0.2	0.305	0.581	0.799	0.939	0.305	0.581	0.799	0.939
0.3	0.294	0.559	0.769	0.905	0.294	0.559	0.770	0.905
0.4	0.275	0.524	0.721	0.847	0.276	0.525	0.723	0.849
0.5	0.219	0.416	0.572	0.672	0.221	0.419	0.577	0.679
Error: max $\mid v_i^n - u_i^n \mid = 0.006$ when $h = 0.2$, $\tau = 0.1$								

6.3   The Method of Lines for the Wave Equation

Let us consider the wave equation (6.1) with the initial boundary conditions (6.2) and (6.3).

$$\frac{\partial^2 u(t,x)}{\partial^2 t} = \frac{\partial^2 u(t,x)}{\partial x^2} + f(t,x), \quad 0 \le x \le l, \ t \ge 0, \tag{6.54}$$

with the initial value conditions

$$u(0,x) = \phi_0(x), \quad \frac{\partial u(0,x)}{\partial t} = \phi_1(x), \quad 0 \le x \le l,$$

and with the boundary value conditions

$$u(t,0) = \nu_0(t), \quad u(t,l) = \nu_N(t), \quad t \ge 0.$$

Substituting into equation (6.1)

$$\frac{\partial^2 u(t,x_i)}{\partial x^2} = \Lambda u_i(t) - \frac{h^2}{12} \frac{\partial^4 u(t,\xi_i)}{\partial x^4}, \quad \xi_i \in (x_{i-1}, x_{i+1}),$$

we obtain the following semi discrete scheme

$$\frac{d^2 u_i(t)}{dt^2} = \Lambda u_i(t) + f_i(t) + E(h,t), \quad i = 1,2,...,N-1, \ t \ge 0, \tag{6.55}$$

where

$$\Lambda u_i(t) = \frac{u_{i-1}(t) - 2u_i(t) + u_{i+1}(t)}{h^2}, \quad u_i(t) = u(t,x_i), \ x_i = ih, \ Nh = l.$$

and the local truncation error

$$E(h,t) = -\frac{h^2}{12} \frac{\partial^4 u(t,\xi_i)}{\partial x^4}.$$

Canceling the truncation error, we get the system of ordinary differential equations

$$\frac{d^2 v_i(t)}{dt^2} = \Lambda v_i(t) + f_i(t), \quad i = 1,2,...,N-1. \tag{6.56}$$

with the initial condition

$$v_i(0) = \phi(ih), \quad \frac{dv_i(0)}{dt} = \phi_1(ih), \quad i = 1, 2, ..., N - 1.$$

Clearly, by boundary conditions, we have

$$v_0(t) = \nu_0(t), \quad v_N(t) = \nu_N(t), \quad t \geq 0.$$

6.3.1   Solution of homogeneous equations

Now, let us solve the system of ordinary differential equations (6.56), by the Fourier series method in the case when $\nu_0(t) = 0$, $\nu_N(t) = 0$ and $f(t, x) = 0$. Substituting into (6.56)

$$v_i(t) = T(t)V_i, \quad i = 1, 2, ..., N - 1, \ t \geq 0,$$

we obtain the following equations for $T(t)$ and V_i:

$$\frac{1}{T}\frac{d^2T}{dt^2} = -\lambda, \quad t \geq 0, \tag{6.57}$$

and

$$-\Lambda V_i = \lambda V_i, \quad i = 1, 2, ..., N - 1, \quad V_0 = 0, \quad V_N = 0. \tag{6.58}$$

As we know, the solutions of the difference equation (6.58) are eigenvalues and the eigenfunctions of the operator $-\Lambda$, so that

$$\lambda_k = \frac{4}{h^2}\sin^2\frac{k\pi h}{2l},$$

and

$$v_i^{(k)} = \sqrt{\frac{2}{l}}\sin\frac{k\pi ih}{l}, \quad i = 0, 1, ..., N, \quad k = 1, 2, ..., N - 1.$$

Integrating the equation (6.57), we find

$$T_k(t) = A_k\cos\sigma_k t + B_k\sin\sigma_k t, \quad k = 1, 2, ..., N - 1, \quad t \geq 0. \tag{6.59}$$

Hence, the general solution of the equation (6.56) is:

$$v_i(t) = \sqrt{\frac{2}{l}}\sum_{k=1}^{N-1}\sin\frac{k\pi ih}{l}A_k\cos\sigma_k t + B_k\sin\sigma_k t, \tag{6.60}$$

where A_k, B_k $k = 1, 2, ..., N - 1$, are coefficients of the Fourier's series for the functions $\phi_0(ih)$ and $\phi_1(ih)$ given in the initial conditions, that is

$$v_i(0) = \phi(ih) = \sqrt{\frac{2}{l}}\sum_{k=1}^{N-1}A_k\sin\frac{k\pi ih}{l}.$$

$$\frac{dv_i(0)}{dt} = \phi_1(ih) = \sqrt{\frac{2}{l}} \sum_{k=1}^{N-1} B_k \sin \frac{k\pi ih}{l}.$$

and

$$A_k = h\sqrt{\frac{2}{l}} \sum_{i=1}^{N-1} \phi_0(ih) \sin \frac{k\pi ih}{l}, \quad k = 1, 2, ..., N-1.$$

$$B_k = h\sqrt{\frac{2}{l}} \sum_{i=1}^{N-1} \phi_1(ih) \sin \frac{k\pi ih}{l}, \quad k = 1, 2, ..., N-1.$$

Example. Let us solve the heat equation

$$\frac{\partial^2 u}{\partial t^2} = \frac{\partial^2 u}{\partial x^2}, \quad 0 \le x \le 1, \quad t > 0,$$

with the initial value condition

$$u(0, x) = x(1 - x), \quad \frac{\partial u(0, x)}{\partial t} = \sin 2\pi x, \quad 0 \le x \le 1,$$

and with the homogeneous boundary value conditions

$$u(t, 0) = 0, \quad u(t, 1) = 0, \quad t \ge 0.$$

Applying the Fourier series method, we can find the theoretical solution

$$u(t, x) = \sqrt{2} \sum_{k=1}^{\infty} c_k e^{-\sigma_k t} \sin k\pi x,$$

where $\sigma_k = \pi^2 k^2$, and the Fourier's coefficients

$$c_k = \sqrt{2} \int_0^1 x(1-x) \sin k\pi x \, dx = \frac{4\sqrt{2}}{\pi^3 k^3}[1 - (-1)^k]. \tag{6.61}$$

By formula (6.60), the approximate solution

$$v_i(t) = \sqrt{2} \sum_{k=1}^{N-1} C_k e^{-\lambda_k t} \sin k\pi ih, \quad i = 1, 2, ..., N-1, \quad t > 0,$$

where the Fourier's coefficients

$$C_k = h\sqrt{2} \sum_{i=1}^{N-1} ih(1 - ih) \sin k\pi ih, \quad k = 1, 2, ..., N-1.$$

These coefficients are calculated by numerical integration of the integrals that appear in formula (6.61), so that

$$C_k \approx c_k = \frac{4\sqrt{2}}{\pi^3 k^3}[1 - (-1)^k], \quad k = 1, 2, ..., N-1.$$

Since $\lambda_k \approx \sigma_k$, therefore $v_i(t) \approx u(t, ih), \quad i = 1, 2, ..., N-1, \quad t \geq 0.$

Solution of the non-homogeneous equations. Let us consider non-homogeneous heat equation in the case when the function $f(t, x)$ has the Fourier sine series representation

$$f(t, x) = \sqrt{\frac{2}{l}} \sum_{k=1}^{\infty} f_k(t) \sin \frac{k\pi x}{l}, \quad 0 \leq x \leq l, \quad t \geq 0,$$

with the Fourier's coefficients

$$f_k(t) = \sqrt{\frac{2}{l}} \int_0^l f(t, x) \sin \frac{k\pi x}{l} dx, \quad k = 1, 2, ...$$

Then, the theoretical solution can be found in the following form:

$$u(t, x) = \sqrt{\frac{2}{l}} \sum_{k=1}^{\infty} T_k(t) \sin \frac{k\pi x}{l},$$

when $u(t, 0) = 0$ and $u(t, l) = 0, \; t \geq 0.$

Indeed, substituting the sine series into (6.54), we obtain the ordinary differential equation

$$\sum_{k=1}^{\infty} [T_k''(t) + \sigma_k T_k(t) - f_k(t)] \sin \frac{k\pi x}{l} = 0, \quad \sigma_k = \frac{k^2 \pi^2}{l^2},$$

with the initial condition

$$\sqrt{\frac{2}{l}} \sum_{k=1}^{\infty} T_k(0) \sin \frac{k\pi x}{l} = \phi(x), \quad 0 \leq x \leq l.$$

This equation generates the decoupled system of equations

$$T_k''(t) + \sigma_k T_k(t) = f_k(t), \quad k = 1, 2, ..., \quad t \geq 0.$$

with the initial conditions

$$T_k(0) = \sqrt{\frac{2}{l}} \int_0^l \phi(x) \sin \frac{k\pi x}{l} dx, \quad k = 1, 2, ...,$$

Using the integrating factor, we find the following solution:

$$T_k(t) = T_k(0)e^{-\sigma_k t} + \int_0^t e^{-\sigma_k(t-\tau)} f_k(\tau) d\tau.$$

Thus, the theoretical solution of the non-homogeneous heat equation is:

$$u(t, x) = \sqrt{\frac{2}{l}} \sum_{k=1}^{\infty} [T_k(0)e^{-\sigma_k t} \sin \frac{k\pi x}{l} + \sin \frac{k\pi x}{l} \int_0^t e^{-\sigma_k(t-\tau)} f_k(\tau) d\tau].$$

Now, let us substitute

$$v_i(t) = \sqrt{\frac{2}{l}} \sum_{k=1}^{\infty} T_k(t) \sin \frac{k\pi ih}{l},$$

into equations (6.56). Then, following the Fourier series method, we obtain the decoupled system of equations

$$T_k''(t) + \lambda_k T_k(t) = f_k(t), \quad k = 1, 2, ..., \quad t \geq 0.$$

with $\lambda_k = \dfrac{4}{h^2} \sin^2 \dfrac{k\pi h}{2l}$, and with the initial conditions

$$T_k(0) = \sqrt{\frac{2}{l}} \int_0^l \phi(x) \sin \frac{k\pi ih}{l} dx, \quad k = 1, 2, ...,$$

Also, by use of the integrating factor, we find

$$T_k(t) = T_k(0)e^{-\lambda_k t} + \int_0^t e^{-\lambda_k(t-\tau)} f_k(\tau) d\tau.$$

Hence, the approximate solution of the non-homogeneous heat equation is

$$v_i(t) = \sqrt{\frac{2}{l}} \sum_{k=1}^{\infty} [T_k(0)e^{-\lambda_k t} \sin \frac{k\pi ih}{l} + \sin \frac{k\pi ih}{l} \int_0^t e^{-\lambda_k(t-\tau)} f_k(\tau) d\tau],$$

for $i = 1, 2, ..., N - 1$.

Example. Let us solve the initial-boundary problem

$$\frac{\partial u(t, x)}{\partial t} = \frac{\partial^2 u(t, x)}{\partial x^2} + \sin 3\pi x, \quad 0 \leq x \leq 1, \ t \geq 0,$$

$$u(0, x) = x(1 - x), \quad 0 \leq x \leq 1,$$

$$u(t, 0) = 0, \quad u(t, l) = 0, \quad t \geq 0.$$

For $f(t, x) = \sin 3\pi x$ and $\phi(x) = x(1 - x)$, $0 \leq x \leq 1$, $t \geq 0$, we calculate

$$f_k(t) = 2 \int_0^1 \sin k\pi x \sin 3\pi x \, dx = \begin{cases} 1, & k = 3, \\ 0, & k \neq 3, \end{cases}$$

$$T_k(0) = \sqrt{2} \int_0^1 x(1 - x) \sin k\pi x dx = \frac{2\sqrt{2}}{\pi^3} \frac{[1 - (-1)^k]}{k^3},$$

$$T_k(t) = \frac{4}{\pi^3} \sum_{k=1}^{\infty} \frac{[1 - (-1)^k]}{k^3} e^{\sigma_k t} + \frac{1}{\sigma_3}[1 - e^{-\sigma_3 t}].$$

Hence, the theoretical solution

$$u(t, x) = \frac{4}{\pi^3} \sum_{k=1}^{\infty} \frac{[1 - (-1)^k]}{k^3} e^{-\sigma_k t} \sin k\pi x + \frac{1}{\sigma_3}[1 - e^{-\sigma_3 t}] \sin 3\pi x.$$

In a similar way, we find the approximate solution

$$v_i(t) = \sqrt{2} \sum_{k=1}^{N-1} C_k e^{-\lambda_k t} \sin k\pi ih + \frac{1}{\lambda_3}[1 - e^{-\lambda_3 t}] \sin 3\pi ih, \quad i = 1, 2, ..., N - 1,$$

where the Fourier's coefficients

$$C_k = h\sqrt{2} \sum_{i=1}^{N-1} ih(1 - ih) \sin k\pi ih, \quad k = 1, 2, ..., N - 1.$$

Let us note that $\lambda_k \approx \sigma_k$ and $C_k \approx T_k(0)$, therefore $v_i(t) \approx u(t, ih)$, $i = 1, 2, ..., N = 1$.

6.4   Exercises

Question 6.1 *Solve the following finite difference equation by the method of separation of variables:*

1. *(a)*

$$\Lambda_t v_i^n = \Lambda_x v_i^{n+1}, \quad (n\tau, ih) \in Q_{h,\tau},$$

$$v_i^0 = \sin 2\pi ih, \quad \frac{v_i^1 - v_i^0}{\tau} = 0, \quad i = 0, 1, ..., N, \quad Nh = 1, \quad (6.62)$$

$$v_0^n = 0, \quad v_N^n = 0, \quad n = 0, 1, ..., \frac{T}{\tau},$$

where $Q_{h,\tau} = \{h, 2h, 3h, ..., Nh\} \times \{\tau, 2\tau, 3\tau, ..., T\}$.

   (b) Give an a priori estimate of the solution v in the norm of the Hilbert space H_h^0.

Question 6.2 *Consider the wave equation*

$$\frac{\partial^2 u(t, x)}{\partial t^2} = \frac{\partial^2 u(t, x)}{\partial x^2} + t\pi^2 \sin \pi x, \quad t \geq 0, \quad 0 \leq x \leq 1,$$

with the initial conditions

$$u(0, x) = 0, \quad \frac{\partial u(0, x)}{\partial t} = \sin \pi x, \quad 0 \leq x \leq 1,$$

and with the boundary conditions

$$u(t, 0) = 0, \quad u(t, 1) = 0, \quad t \geq 0,$$

1. *(a) Approximate the initial-boundary problem by the finite difference with weight and determine the local truncation error.*

(b) Solve the finite difference scheme with weight by the method of separation of variables and estimate the global error of the method.

(c) Solve the finite difference scheme with weight $\sigma = 1$ using the module `hyperbolicWeghtEquation`.

Question 6.3 Use the `hyperbolicWeightEquation` to solve the following initial boundary problem

$$\frac{\partial^2 u(t, x)}{\partial t^2} = \frac{\partial^2 u(t, x)}{\partial x^2} + e^{-t}(2 + x - x^2), \quad t \geq 0, \quad 0 \leq x \leq 1,$$

with the initial conditions

$$u(0, x) = x(1 - x), \quad \frac{\partial u(0, x)}{\partial t} = x(x - 1),$$

and with the homogeneous boundary conditions

$$u(t, 0) = 0, \quad u(t, 1) = 0,$$

Question 6.4 Consider the following initial boundary problem:

$$\frac{\partial^2 u(t, x)}{\partial t^2} = \frac{\partial^2 u(t, x)}{\partial x^2}, \quad 0 \leq x \leq 2, \ t \geq 0,$$

$$u(0, x) = x(1 - x), \quad \frac{\partial u(0, x)}{\partial t} = \sin \frac{\pi x}{2}, \quad 0 \leq x \leq 2,$$

$$u(t, 0) = 0, \quad u(t, 2) = 0, \quad t \geq 0.$$

1. (a) Find the theoretical solution $u(t, x)$ using the Fourier series method.

(b) Solve the initial boundary problem by the method of lines.

(c) Give an estimate of the error of the method of lines.

Question 6.5 Prove $O(h^2)$ convergence of the method of lines in the norm of the Hilbert space H_h^0 when it is applied to the wave equation with the initial-boundary conditions.

References

1. Ames, W. F., (1992), Numerical Methods for Partial Differential Equations, Academic Press, INC.

2. Aubin, J. P., (1972), Approximation of Elliptic Boundary Value Problems, Wiley- Interscience

3. Berezin, I.S. & Zidkov, N.P. (1962), Numerical Methods, v.II,Moskaw

4. Boyce, W.E. & DiPrima, R.C., (1992), Elementary Differential Equations, John Wiley & Sons, Inc., New York.

5. Bramble, I.H. & Hubbard, B.E. (1964), On finite difference analogue of an elliptic boundary problem which is neither diagonally dominant nor of non-negative type, J. Math. Phys, 43, p. 117

6. Collatz, L., (1936), Angew. Math. Mech., 16, p.239.

7. Euler, L. (1768), Institutiones Calculi Integralis, Petersburg, (See Leonardi Euleri Opera Omnia, Ser. I, v.XI. p.424, Taubner Verlag Leipzig, 1913).

8. Fatula, S.O., (1988), Numerical Methods for Initial Value Problems in Ordinary Differential Equations, Academic Press, INC., New York.

9. Forsythe, G. E., & Wasow, W. R.,(1960), Finite difference Methods for Partial Differential Equations. John Wiley,

10. Gear, C.W. (1971), Numerical Initial Value Problem in Ordinary Differential Equations, Prentice-Hall INC.

11. Gerschgorin, S. Z.,(1930), Angew. Math. Mech., 10, p.373.

12. Hall, C.A. (1984), Numerical Analysis of Partial Differential Equations. Hemisphere Publishing Company.

13. Henrici, P. (1962), Discrete Variable Methods in Ordinary Differential Equations, John Wiley & Sons.

14. Lambert, J.D.(1973), Computational Methods in Ordinary Differential Equations, John Wiley & Sons.

15. Michlin, S.G. & Smolnicky C.L. (1963), Approximate Methods for Solving Differential and Integral Equations, Nauka, Moscow.

16. Milne-Thomson, L.M. (1981), Calculus of Finite Differences, Chelseta Publishing Company, New York.

17. Mickens, R.E. (1987), Difference Equations, Van Nostrand Reinhold Company.

18. Mikeladze, Sh., (1941), Izv. Acad. Nauk, SSSR, 5. p. 57.

19. Mitchell, A.R. & Griffiths D.F. (1975), The Finite Difference Method in Partial Differential Equations, John Wiley & Sons, New York.

20. Quarteroni, A., & Valli, A., (1994), Numerical Approximation of Partial Differential Equations, Springer-Verag, New York.

21. Richardson, L.F.,(1910), Trans. R. Soc., London, A210, p. 307.

22. Richtmyer, R. D., (1967), Difference Methods for Initial Value Problems, 2nd ed. Wiley Interscience, New York.

23. Runge, C.Z., (1908), Math. Phys., 56, p.225.

24. Lapidus, L. and Pinder, G.F. (1982) Numerical Solution of Differential Equations, John Wiley

25. Samarski, A. A. (1971), Introduction to Finite Difference Schemes, Nauka, Moscow.

26. Shampine, L. F., (1994), Numerical Solution of Ordinary Differential Equations, Chapman & Hall,

27. Smith, G.D. (1985), Numerical Solution of Partial Differential Equations: Finite Difference Methods, Clarendon Press-Oxford.

28. Stoer, J. & Bulrisch R. (1980), Introduction to Numerical Analysis, Springer-Verlag, (1980).

29. Styš, T. (1980), A Discrete Maximum Principle, Dissertationes Mathematicae, CLXXXII, Warsaw.

30. K. Styš & T. Styš,(1997), An Optimal Algorithm for Certain Boundary Value Problem, Journal of Computational and Applied Mathematics, v. 83, pp. 195-203.

31. Varga, R.S. (1966), On a maximum principle, SIAM J.Numer. Anal. 3.

32. Wakulicz, A. ((1970), Methodi Alle Differenze per Equazioni Differentiali alle Derivate parziali, Roma.

33. Wolfram, S., (1992),Mathematica a System for Doing Mathematics by Computer, Addison-Wesley Publishing Company, New York.

INDEX

www.ingramcontent.com/pod-product-compliance
Lightning Source LLC
Chambersburg PA
CBHW041710210326
41598CB00007B/603